Quantum Mechanics in the Single-Photon Laboratory (Second Edition)

Online at: https://doi.org/10.1088/978-0-7503-6315-0

IOP Series in Quantum Technology

Series Editor: **Barry Garraway** (School of Mathematical and Physical Sciences, University of Sussex, UK), **Barry Sanders** (Institute for Quantum Science and Technology, University of Calgary, Canada) and **Lincoln Carr** (Quantum Engineering Program, Colorado School of Mines, USA)

About the series

The IOP Series in Quantum Technology is dedicated to bringing together the most up to date texts and reference books from across the emerging field of quantum science and its technological applications. Prepared by leading experts, the series is intended for graduate students and researchers either already working in or intending to enter the field. The series seeks (but is not restricted to) publications in the following topics:

- Quantum biology
- Quantum communication
- Quantum computation
- Quantum control
- Quantum cryptography
- Quantum engineering
- Quantum machine learning and intelligence
- Quantum materials
- Quantum metrology
- Quantum optics
- Quantum sensing
- Quantum simulation
- Quantum software, algorithms and code
- Quantum thermodynamics
- Hybrid quantum systems

A full list of titles published in this series can be found here: https://iopscience.iop.org/bookListInfo/iop-series-in-quantum-technology.

Quantum Mechanics in the Single-Photon Laboratory (Second Edition)

Muhammad Sabieh Anwar
Department of Physics, Syed Babar Ali School of Science and Engineering, Lahore University of Management Sciences (LUMS), Lahore, Pakistan

Faizan-e-Ilahi
Department of Physics, Syed Babar Ali School of Science and Engineering, Lahore University of Management Sciences (LUMS), Lahore, Pakistan

Syed Bilal Hyder
Department of Physics, Syed Babar Ali School of Science and Engineering, Lahore University of Management Sciences (LUMS), Lahore, Pakistan

Muhammad Hamza Waseem
Department of Physics, Syed Babar Ali School of Science and Engineering, Lahore University of Management Sciences (LUMS), Lahore, Pakistan

IOP Publishing, Bristol, UK

Muhammad Sabieh Anwar, Faizan-e-Ilahi, Syed Bilal Hyder and Muhammad Hamza Waseem have asserted their right to be identified as the authors of this work in accordance with sections 77 and 78 of the Copyright, Designs and Patents Act 1988.

ISBN   978-0-7503-6315-0 (ebook)
ISBN   978-0-7503-6313-6 (print)
ISBN   978-0-7503-6316-7 (myPrint)
ISBN   978-0-7503-6314-3 (mobi)

DOI   10.1088/978-0-7503-6315-0

Version: 20240801

IOP ebooks

British Library Cataloguing-in-Publication Data: A catalogue record for this book is available from the British Library.

Published by IOP Publishing, wholly owned by The Institute of Physics, London

IOP Publishing, No.2 The Distillery, Glassfields, Avon Street, Bristol, BS2 0GR, UK

US Office: IOP Publishing, Inc., 190 North Independence Mall West, Suite 601, Philadelphia, PA 19106, USA

You bring another tank, we will bring our stone.

منكم دبابة أخرى ومنا حجر

—**Mahmoud Darwish** *(1941–2008)*

Contents

Preface

For many years I had dreamt of making a laboratory for our students that could demonstrate fundamental facets of quantum physics, quantum information, and quantum computing. Although our students at PhysLab, Lahore University of Management Sciences (LUMS), Syed Babar Ali School of Science and Engineering do work with superconducting quantum interference devcies (SQUIDs), Franck–Hertz tubes, and lasers, I was looking for ideas that were more 'grainy', counter-intuitive, and quantitative, and which could directly relate to quantum interference, entanglement, density matrices, nonlocality, and reveal the eerie aspects of quantum physics.

Luckily, I came across two students, Hamza Waseem and Faizan-e-Ilahi, who, being students of electrical engineering at the University of Engineering and Technology (UET), Lahore, converted their deep sense of deprivation of a formal education in physics to a craving for experimenting with single photons. I completed my electrical engineering from UET, and so I have a fondness for this institution. Hamza is now pursuing a doctorate in Physics from the University of Oxford (where I studied 15 years ago), while Faizan continues as an independent researcher. After a year and a half of dedicated effort, struggling between coursework at their parent institution and the tough challenges we had laid out for ourselves, Hamza and Faizan managed to complete a suite of fundamental experiments on quantum mechanics. This book is about these experiments. It not only describes the experiments but also provides all the necessary working knowledge of quantum mechanics, quantum states, and quantum operators that is required to motivate the reader toward performing these experiments, and providing the necessary tools to interpret and understand the experimental outcomes.

At some stage in the project, our undergraduate student at LUMS, Bilal, also joined the group and expanded the scope of many experiments, and provided the motivation to improve the script and add more background material. Hence, what is in front of you is the book's second and considerably expanded edition, with a new experiment on nonlocal quantum erasure, the rectification of errors that had crept into the first edition, and improved explanations on the various facets of single-photon experiments.

This book starts with a survey of how research laboratories around the world have helped to create a portfolio of some wonderful experiments that can be easily translated for instructional purposes in physics teaching laboratories worldwide. As such, these experiments are a true distillation of cutting-edge research in quantum optics and quantum information science with single photons and epitomize how research in the laboratory can directly enrich teaching of a counter-intuitive mathematical and physical framework that we call the quantum theory. Since our working units for qubits are photon's polarization state, we make a brief digression into the classical picture of light. Some of the mathematical tools introduced in the second chapter carry over, in a quantum garb, to the quantum experiments. We also describe some experiments with lasers that can act as a springboard for the quantum experiments.

Enter the quantum description of nature with states of qubits living on the Bloch sphere. The third chapter builds just enough of the mathematical description that is required to understand how we explain quantum states, how they change with time or when acted upon by certain transforming elements, and finally how we measure them. Connections with photon polarization are built as we move along. This chapter also provides us the opportunity to introduce quantum entanglement, which is hailed as the quantum world's most closely guarded and most cherished secret.

The background material which spans over three chapters is not a substitute to a standard semester long course on quantum mechanics, but does provide enough material to the non-physics major, or engineer, or even the general science student, to understand how quantum mechanics unfolds in reality, inside a laboratory, and reveals itself in full glory with simple 'but not simpler' instruments and gadgets.

Chapters four through eight describe the experiments. We have made a conscious effort in classifying these experiments into three categories: those that deal with the 'quantumness' of these quantum experiments, those that demonstrate how the quantum reality is neither local nor realist, and finally how we can measure the quantum state. All of these various kinds of experiments highlight one aspect or the other of quantum reality.

None of these experiments are novel. They have been tried many times by many people around the world. Several excellent laboratory manuals and pedagogical articles recount these experiments. We have benefited from all of these prior demonstrations. In particular, we have drawn upon the experiments described in Mark Beck's textbook 'Quantum Mechanics: Theory and Experiment' and these experiments are mostly an explanation of Beck's innovation. However, this book is not fashioned as a laboratory manual. A smart professor with some expertise in practical optics can easily decipher our narrative and reconstruct these experiments, but the text is clearly not written as a laboratory recipe book for the student. The style is more like a textbook, but a textbook that describes real experiments. A close analogy is with the 'Experiments in Modern Physics' by A Melissinos and J Napolitino.

I would really like to say that the experiments outlined in this book are truly multidisciplinary. Even though the motivation is understanding and revealing the quantum nature of fields, they truly make the experimenter feel what it is like in a real research environment, including aligning optical beams, searching for the bull's eye shot of a laser focus, chasing nanosecond long electronic pulses on an oscilloscope screen, making electronic circuits that can change the voltage levels on high speed signals, computer programming, and connecting experiment with an unusual theoretical structure, followed by data analysis.

We attempt to present our single-photon quantum experiments in a modular fashion, with one building on the other. The concluding chapter also presents some suggestions for teachers and researchers who are interested in reproducing these experiments; for possible pathways of building these activities in our advanced physics laboratories; or for offering these experiments, in a simple to complicated ladder, to students undertaking laboratory courses. The appendices provide some useful technical details about the basics and programming of field-programmable gate arrays, and also spells out the inventory that we employed in our work. The inventory is by no means comprehensive or an endorsement of one instrument over the other—it merely reflects what worked best for us, and in most cases, in the first go. The inventory and datasets can also be downloaded from our website http://www.physlab.org/qmlab.

Muhammad Sabieh Anwar
Lahore, Pakistan, June, 2024

Acknowledgments

By no means are we alone in this effort. This initiative is a pedagogical partnership and is inspired by several colleagues, most of them in the United States, who go to extreme lengths to making the teaching of quantum physics simple, intuitive, elegant, and above all a moment of joy. I would like to thank Drs Mark Beck, Enrique Galvez, Richard Haskell, Theresa Lynn, Art Hobson, Suhail Zubairy, and David Van Baak for inspiring me, through their open-source books, personal training, and published articles, for carrying out the mission of using physics as an agent of change and well-being. Haskell and Lynn at the Harvey Mudd College led a laboratory immersion program on single photons in 2017 that helped me to learn many tools of the trade. Thanks to the ALPhA initiative for organizing these sessions.

After completing my formal studies, I joined the Lahore University of Management Sciences (LUMS), a not-for-profit University in Pakistan as an assistant professor. The objective that lay in front of us in 2007 was setting up a microcosm of science-infused teaching and research ecosystem in the country that would become a regional role model in education and discovery. I must say that I was really lucky. The university provided me, and all of my colleagues, the academic freedom that allowed us to curate new laboratories, design innovative multidisciplinary courses, seed research areas that were non-existent in the country, and incite a generation of holistically trained science and engineering graduates who could 'think globally' but 'act locally'. There were many inspirers to this work, notably Prof Asad Abidi at the University of California Los Angeles who in fact brought me to Pakistan and gave me the creative space, with an open heart, and with the utmost degree of elegance and nuanced encouragement helped me build the PhysLab.

In this environment, I got the chance to cultivate a laboratory called the PhysLab, (http://www.physlab.org), which was meant to create a national and regional platform for research-inspired instruction in experimental physics with all its resources and blueprints made public, so that other institutions around us could also benefit from any humble work that we did. This laboratory showcases a tapestry of experiments in basic and advanced physics, ranging from fluids to cosmic ray muons, and pendulums to superconductors.

In this overarching framework, the single-photon laboratory narrated in this book is our effort to share how quantum physics can unfold in the real world with demonstrable experiments, yielding quantitative data which can be easily analyzed revealing insightful interpretations of quantum mechanical postulates.

The PhysLab depends on the unwavering support of a host of remarkable individuals and I cannot mention each one of them, but still I cannot miss Sohaib Shamim (late), Amrozia Shaheen, Junaid Alam, Muhammad Rizwan, Dr Umar Hassan in Rutgers University, Dr Ali Akbar in the University of Nottingham, Abdullah Ijaz, Umar Hassan, Hassaan Majeed, Khadim Mehmood, Muhammad Shafique, Azeem Iqbal, Kaniz Amna, Muzammil Shah, Abdullah Irfan, Ayyaz Mehmood, and Ali Hassan. I cannot thank my department heads enough,

Dr Muhammad Faryad and Dr Adam Zaman, for allowing me to pursue my passions and Arshad Maral, the department's secretary, to facilitating me at every level. LUMS has also generously provided me with financial support.

The eighth chapter of this book on quantum state tomography is a joint work with my PhD student, Ali Akbar, and certain extensions of this work are part of his PhD dissertation. Ali, Alamdar, and Obaidullah have immensely helped me in setting up the Ibn-e-Sahl Optics Lab (https://www.physlab.org/optics-lab/), which became the nursery for these single-photon quantum experiments.

Finally, gratitude comes to a different level when your loved ones support you in invisible ways. This book was purely the idea of my father, Dr Saadat Anwar Siddiqi, who has stood behind me in my career as a strong inspirer. My father passed away when the first edition of the book was published. I really owe it to him and cannot stop missing his company at any moment of my life. My mother Shahida Bano has given me the prayers to carry on and define new targets for me every time. She keeps us all together. My wife Hina Zulfiqar and my children, Fatimah and Khadeejah, have also given away parts of their lives in letting me engage in these academic activities.

I especially thank the staff of the Institute of Physics Publishing, particularly John Navas, Phoebe Hooper, and Sarah Armstrong, and other members of the editorial and production teams who are invisible to us, for their timely support, suggestions, and meticulous care for this project. It has been a true pleasure working with IOP.

Muhammad Sabieh Anwar
June, 2024

We thank Dr Sidra Farid, Associate Professor of Electrical Engineering at University of Engineering and Technology (UET), Lahore, for her support and encouragement to pursue this project. We are grateful to Mah Para Iqbal and Zahra Tariq for their help in performing preliminary experiments and designing figures. Thanks are also due to the technical and support staff at LUMS.

Most importantly, we would like to express our immense gratitude to Dr Sabieh Anwar for letting us work in his laboratory and keeping us on our toes. The subject matter of this book signifies our transition from engineering to physics, and is therefore especially close to our hearts. It has truly been a great privilege working at PhysLab.

Faizan-e-Ilahi
April, 2024

First and foremost, I would like to thank Dr Sabieh Anwar for giving me the opportunity to work in the single-photon laboratory and bringing me on board for this project. It was with his guidance, support, and trust that we were able to complete this project successfully. I would also like to express my gratitude to the Physlab team, who were readily available to assist us with everything we needed.

Among the team members, Ali Hassan, Umar Hassan, Muhammad Shafique, Ayaz Mahmood, and Khadim Mahmood deserve a special mention.

Most importantly, I want to express my gratitude to my family. My grandfather Muzammil's determination and quest for knowledge have been a great inspiration to me. My parents, Nasarullah and Farkhanda, and my siblings, Abdullah, Muneeza, and Abeeha, have supported me through all my busy schedules and ambitious goals. They have been there for me through every step of the journey.

I also want to acknowledge my peers and friends who supported me throughout the duration of this project and stuck by me, even when I had to cancel many plans due to work commitments. I extend my gratitude to Mahid Anjum, Wardah Mahmood, Abdullah Ijaz, Mariam Arif, Abdullah Atif, Saleha Shahid, Nouman Shahzad, Saira Masood, Haider Khan, Fida Hussain, Shakeel Afzal, Ahmad Ali, Moazzan Waheed, and Muizz Butt for their unwavering support throughout this journey.

Syed Bilal Hyder
April, 2024

I am immensely grateful to Prof Sabieh Anwar for giving me the opportunity to join the single-photon lab project in 2018. For an engineering student longing to delve into fundamental physics, this project, along with the ensuing book, provided precisely that opportunity and ultimately served as a springboard for a career in physics.

Parts of this book were written after I started my graduate studies at the University of Oxford. I would like to acknowledge the financial support for my graduate studies from the Rhodes Trust and Magdalen College, Oxford.

The previous edition of this book served as a supplementary text for the graduate Quantum Information course at Oxford. I would like to express my gratitude to Prof John Gregg for referring this book to the students of Magdalen College. I also thank my students and colleagues for constantly challenging my understanding of quantum physics.

I express my heartfelt appreciation to my supervisors, Prof Bob Coecke and Dr Alexy Karenowska, for their unwavering encouragement and support throughout my time at Oxford.

Above all, I wish to extend my deepest gratitude to my family and friends for their emotional support and belief in my dreams and aspirations.

Muhammad Hamza Waseem
April, 2024

Author biographies

Dr Muhammad Sabieh Anwar

Dr Muhammad Sabieh Anwar is a professor of physics at the Lahore University of Management Sciences (LUMS), Syed Babar Ali School of Science and Engineering, Pakistan. Ideas from his physics instructional laboratories have been replicated in about 10 Pakistani universities. His laboratories and research are presented on https://www.physlab.org. He loves teaching physics with hand crafted in-class demonstrations and some of his courses on modern physics, electromagnetism, magnetism, and quantum physics can be seen on YouTube. His research interests encompass spintronics, magnetism, and optics. Sabieh is also the general secretary of the Khwarizmi Science Society (https://www.khwarizmi.org), which is aimed at popularization of science at the grassroots levels in Pakistan. He leads the iconic science festival called the Lahore Science Mela (https://www.ksslsm.org), which attracts hundreds of thousands of keen visitors from across the region. Prior to joining LUMS in 2007, Sabieh was a post-doc in chemistry and materials science at the University of California, Berkeley and a PhD student, as Rhodes Scholar, at Oxford University.

Faizan-e-Ilahi

Faizan-e-Ilahi is a student of physics and is captivated by the intricacies of quantum mechanics. He is inspired by the profound insights of the renowned physicist, Leonard Susskind, whose works ignited his passion for quantum mechanics. During his undergraduate studies, he, along with some amazing friends, embarked on the journey to build first single-photon quantum mechanics laboratory of Pakistan under the supervision of Dr Muhammad Sabieh Anwar. Beyond science, he finds solace in Urdu and Persian literature. He is particularly drawn to the imaginative and philosophical poetry of Mirza Ghalib.

Syed Bilal Hyder

Syed Bilal Hyder earned his BS in Physics from LUMS Syed Babar Ali School of Science and Engineering as an NOP scholar. After graduating, he joined PhysLab (https://www.physlab.org) as a full-time researcher and undertook multiple projects to upgrade the single-photon laboratory. His research interests lie in quantum communication and quantum computing architecture. Besides physics, he shares a passion for computer science and is also a part ofEducative (https://www.educative.io), where he helped develop the 'Learn to Code' skill path. Moreover, Bilal's commitment extends to science outreach. He is an active science communicator, and led science demonstrations and instrument building for Lahore Science Mela in 2022 and 2023.

Muhammad Hamza Waseem

Muhammad Hamza Waseem is pursuing a DPhil in Physics at the University of Oxford, funded by the Rhodes Trust and Magdalen College. His current research spans quantum foundations, quantum science education, applied category theory, and mathematical linguistics. Passionate about science popularization, Hamza has contributed to outreach events in Pakistan—notably the Lahore Science Mela—and in the UK, such as Royal Institution Masterclasses, Pint of Science Festival, and Oxford Maths Festival. He works as a research scientist at Quantinuum and teaches mathematics and physics at Oxford. He is also helping to develop 'Quantum in Pictures,' a novel framework, graphical language, and curriculum aimed at making quantum physics accessible to a wider audience.

List of abbreviations

APD	Avalanche Photodiode
BBO	β-Barium Borate
BDP	Beam Displacing Polarizer
BS	Beam Splitter
CCU	Coincidence Counting Unit
CHSH	Clauser–Horne–Shimony–Holt
EPR	Einstein–Podolsky–Rosen
FPGA	Field-Programmable Gate Array
HWP	Half-wave Plate
PBS	Polarizing Beam Splitter
PI	Polarization Interferometer
PC	Personal Computer
QWP	Quarter-wave Plate
SPCM	Single-Photon Counting Module
SPDC	Spontaneous Parametric Downconversion
TTL	Transistor-transistor Logic
HDL	Hardware Description Language
UART	universal Asynchronous Receiver/Transmitter

List of quantum optics experiments

Q1   Spontaneous Parametric Downconversion
Q2   Testing the Particle-like behavior of Light
Q3   Estimating the Polarization State of Single Photons
Q4   Visualizing the Polarization State of Single Photons
Q5   Single-Photon Interference and Quantum Eraser
Q+NL   Nonlocal Quantum Eraser
NL1   Freedman's Test of Local Realism
NL2   Hardy's Test of Local Realism
NL3   CHSH Test of Local Realism
QST   Quantum State Tomography

IOP Publishing

Quantum Mechanics in the Single-Photon Laboratory (Second Edition)

Muhammad Sabieh Anwar, Faizan-e-Ilahi, Syed Bilal Hyder and Muhammad Hamza Waseem

Chapter 1

Introduction

The 2022 Nobel Prize in Physics was awarded to Alain Aspect, John F Clauser, and Anton Zeilinger for their seminal contributions in demonstrating and sealing the debate on a property of quantum systems called entanglement. This topic unifies two disciplines of human scientific intellect and technological ingenuity, namely optics and quantum mechanics.

Optics, the study of light, is arguably one of the oldest branches of natural sciences. In contrast, quantum mechanics was born much later, in the twentieth century. Some label it as the most successful theory in physics ever devised because it correctly explains a vast number of physical phenomena. Nevertheless, scientists had a hard time getting comfortable with quantum mechanics because it shook their classical world view, as well as their intuition. This book deals with many strange ideas of the quantum world and will highlight experimental demonstrations that the readers of this book can choose to embark on building for themselves to see the quantum counter-intuition unfold with all its majestic glory.

According to the celebrated physicist Richard Feynman, 'the only mystery' of quantum mechanics is superposition [1], which has some really astonishing effects. Niels Bohr, in the early part of the previous century declared 'those who are not shocked when they first come across quantum theory cannot possibly have understood it' [2].

To understand superposition, let us consider a system with only two possible states, labeled 0 and 1. These states may correspond to a particle taking one of two possible paths. We can do some measurement and determine which path the particle is opting to travel. There is nothing unusual and surprising about this in the classical scenario. Enter quantum superposition, and we get a bizarre possibility: the system can be in a superposition of both the 0 and 1 states, i.e., the particle is presumably taking both paths at the same time. Quantum mechanics tells us that the system is in superposition as long as we are *unable* to determine which path the particle is taking. As soon as a measurement is performed, the superposition state collapses and the

doi:10.1088/978-0-7503-6315-0ch1

particle is found to take either one of the two paths. Hence, the act of measurement makes the particle choose one path. This creation of choice is something called collapse or projection. Now, there is a specific probability of finding the particle in either of the two paths but it is not possible to know in a confirmatory manner, prior to performing the measurement, which path the particle has actually taken.

This bizarre result has confounded scientific thinkers ever since the inception of quantum mechanics. Even Einstein had serious reservations. Adding to the quandary, in 1935 Einstein along with Podolsky and Rosen [3] published a thought experiment and showed that quantum mechanics proposes 'nonlocal' behavior. Since he believed in localism, Einstein concluded that quantum mechanics must either be wrong or incomplete. This debate, however, gave rise to the property of quantum entanglement.

Let us define quantum entanglement in terms of an optical setup. The nonlinear process of optical spontaneous parametric down-conversion [4] can produce a photon pair that is entangled in polarization. Polarization of light is usually expressed in terms of two mutually independent orthogonal linear states, called the horizontal and the vertical. One example of an entangled state would be the superposition of the state in which photons 1 and 2 are both horizontally polarized, and the state where both photons are vertically polarized. Here, superposition implies the photon pair to be in the two aforementioned states *simultaneously*. Again, the photons in the pair will be entangled as long as no measurement to determine their state is made. If we use a horizontally or vertically oriented polarizer to determine the state of one of the two photons, the measurement collapses the entangled state and the other unmeasured photon is predicted to be in the same polarization state. For example, if photon 1 is measured to be horizontally polarized, then we immediately know that photon 2 is also in horizontal polarization. Hence, measurement of the polarization state of one photon deterministically determines the polarization state of the other photon of the entangled pair. This strange result of quantum mechanics was termed 'spooky action at a distance' by Bromberg [5]. Terms such as 'spooky' are rare in mainstream physics.

The Einstein–Podolsky–Rosen thought experiment where particles are entangled in the momentum degree of freedom remained untested until 1964, when John Bell formulated a set of experimentally testable inequalities that could prove the local or nonlocal nature of reality [6]. 'Bell's inequalities', if verified, would debunk nonlocality and prove quantum mechanics incorrect or incomplete. On the other hand, violation of these inequalities would be a verdict in favor of quantum mechanics' most bizarre prediction: nonlocalism. A series of experiments [7] have indeed shown the unequivocal violation of Bell's inequalities, hence proving that nature is nonlocal and quantum mechanics, as we know, is *indeed* a valid theory.

In the last few years, technological advancement has caused noteworthy progress in investigating and harnessing the nonlocal properties of nature. As a result of these advances, we have quantum teleportation [8], quantum cryptography [9], quantum illumination [10], and quantum interferometric imaging [11]. Work under these themes has given birth to the sister fields of quantum information and quantum communication [12].

The technological 'quantum leap' has also brought quantum mechanical experiments to the tabletop by significantly reducing the size and cost of the apparatus involved. One class of such experiments comprises single photons. These experiments, although minimalistic, are powerful enough to demonstrate the fundamentals of quantum mechanics that have, in the last century, eluded many a great physicist. These single-photon experiments also lie at the heart of cutting edge applications in quantum computing and cryptography [12]. The relative ease to produce pairs of entangled photons has therefore brought us to the juncture where we can do quantum mechanics with relatively simple and affordable optical setups [13].

In particular, a number of experiments incorporating individual photons or correlated photon pairs have been devised for supplementing the undergraduate and postgraduate teaching of quantum mechanics [14, 15]. Many of these experiments have been described in the last two decades. Some of the important experiments include the observation of the quantum nature of light [16], interference of single photons [17], locality tests [7, 13, 18] and quantum erasure [19]. This is a book that describes some of these experiments.

Experiments which 'vindicate' the quantum reality of nature fall into various categories. In this book, we will use single photons as our archetypal quantum bits and focus on their polarization degree of freedom. This starting chapter surveys some of the key experiments and recounts major developments that have led to making these verifications more accessible and approachable to physics educators and students.

Apparently, many modern physics textbooks portray the photoelectric effect as an early example of departure from the classical world. This is true, but only partially. Contrary to popular belief held by many students (and teachers), an explanation of the photoelectric effect does not strictly require the existence of photons [20, 21]. Lamb and Scully showed that the photoelectric effect could be explained using a semiclassical model where the detector atoms are considered quantized but the light is deemed classical [22]. An experiment to prove the existence of photons should rather prove the 'granular' nature of light *itself*. The results of such an experiment should not be explicable through the classical wave theory of light.

Some of these pioneering experiments were in fact described in the articles [23–25]. Later on, in 1986, Grangier *et al* performed a simple experiment proving the particulate nature of light [26]. They observed the absence of coincidences between photo-detections at the transmission and reflection outputs of a beam-splitter. Their results demonstrated that the field incident on the beam-splitter could be described by a single-photon state. Hence, it was experimentally shown that a single particle of light can be detected only once. In a pioneering piece of work, reported in 2004, Thorn *et al* adapted the same experiment for an undergraduate laboratory [16].

Experiments on interference and quantum erasure are closely related. When light is made to pass through an interferometer, the interference pattern visibility is dependent on the 'which-way' information that is available to the observer. If the which-way information is not available or 'erased', high visibility interference fringes are obtained. For example, polarization analysis can be performed using polarizers, which can provide or erase which-way information, determining the visibility of interference fringes [17, 27, 28]. Gogo *et al* [29] demonstrated a quantum eraser through correlated photon pairs. The pump laser beam is split into so-called idler and signal beams. The which-way information is obtained by performing measurements on the idler beam. Other examples of quantum eraser experiments can be found in [30, 31].

Violation of locality in quantum mechanics stems from the idea of the EPR paradox [3]. Bell, through his inequalities, pointed out that violation of locality in quantum mechanics could be experimentally tested [6]. Bell's work motivated similar testable inequalities, which are now collectively termed Bell's inequalities. Almost all of these inequalities verify the foundational correctness of quantum mechanics and indicate that nature indeed violates locality [7]. The first experimental test of a Bell's inequality, proposed by Freedman, was performed in 1972 [32]. The Clauser–Horne–Shimony–Holt (CHSH) inequality—the now-standard version of Bell's inequality—for optical settings was proposed by Clauser *et al* [33] and its violation was demonstrated by Aspect *et al* [34].

Greenberger *et al* pointed out the possibility of an all-or-nothing test of locality [35]. The original argument of Bell was statistical, i.e., it was based on different classical and quantum mechanical predictions of the probability of occurrence of some results. However, in Greenberger's test, classical mechanics predicts a certain event, while quantum mechanics predicts an impossible event. In all of these cases, the experimental results have been demonstrated to agree with quantum mechanics [36].

Furthering this theme forward, in 1993 Hardy derived a version of Bell's inequality that is significantly easier to perform experimentally and comprehend [37]. The experimental tests of Hardy's locality have also been shown to remarkably agree with quantum mechanics [38–40].

Over the last two decades, experimental tests of Freedman's, Hardy's, and CHSH inequalities have been successfully performed strictly within undergraduate laboratory resources [7, 13, 18, 41]. Similar resources have been utilized to perform single-qubit measurement [14], two-qubit quantum state tomography [42], as well as single-qubit process tomography [43]. Table 1.1 lists some important research groups along with their published experiments suitable for exploring quantum optics and quantum information in a teaching laboratory.

This book emanates from a senior year capstone project, wherein we have recreated some of the aforementioned experiments and devised several additions as well. The purpose of this work was to design a series of experiments using single-photon sources and optical setups that demonstrate the fundamental principles of quantum mechanics. We think that these experiments, which are currently housed at PhysLab (http://www.physlab.org), Lahore University of Management Sciences (LUMS), Lahore (Pakistan), can be employed as a laboratory for pedagogical purposes or as a sandbox for experimental research in quantum optics, quantum information processing, and quantum computing.

This book is organized as follows:

- Chapter 2 is a brief discussion of the classical view of light along with three experiments. Exploring classical optics experiments first is beneficial because quantum optics experiments often build upon foundations set by their classical analogs.

Table 1.1. Some research groups with significant experimental work on single-photon-based quantum mechanics.

Research group	Publications
Beck's group at Reed College.	Proof of existence of photons [16, 44], single-photon interference [14], tests of local realism [7, 13, 14], quantum eraser [29], entanglement witness [45], and EPR steering [46].
Galvez's group at Colgate University.	Single-photon interference [17, 47], bi-photon interference [17], time-energy interference [48], quantum eraser [49], a test of local realism [50], and the Hong–Ou–Mandel interferometer [51].
Kwiat's group at the University of Illinois at Urbana-Champaign.	Source of polarization-entangled photons [52], tests of local realism [40, 53], and quantum state tomography [42, 54].
Lukishova's group at the University of Rochester.	Single-photon sources and photon antibunching [55].

- Chapter 3 gives a short overview of quantum theory in terms of photons. This chapter will help the readers develop enough understanding of quantum physics and its nomenclature to understand the experiments performed in this book.
- Chapter 4 furnishes general details of our single-photon laboratory and documents two experiments to *directly* investigate the quantum nature of light.
- Chapter 5 describes experiments to determine the polarization of single-photons. Polarization lies at the heart of all quantum photonics experiments.
- Chapter 6 covers three experiments to test the existence the unique property of quantum entanglement and violation of Bell's inequalities.
- Chapter 7 contains experiments illustrating the peculiar effects, including nonlocality, resulting from the entanglement of particles.
- Chapter 8 discusses quantum state tomography, which is an advanced measurement technique that measures everything that can be measured about quantum particles.

Finally, chapter 9 concludes our work. A detailed chapter for programming of a coincidence counting unit on an FPGA board is also included as appendix A for readers who are interested in building such experiments themselves. All data and supplementary details pertaining this book will be furnished on the book's website:

http://www.physlab.org/qmlab

The optical and optomechanical equipment required for these experiments is readily available from mainstream suppliers such as Thorlabs, Edmund Optics, and Newport. Single-photon detectors can be sourced from Thorlabs, Excellitas, IDQ, and similar companies. We have also included a complete list of components used in these experiments in appendix B. We hope that these lists will be found to be useful resources.

References

[1] Feynman R P, Leighton R B and Sands M 1965 *Lectures on Physics: Quantum Mechanics* **vol III** (Reading, MA: Addison-Wesley)

[2] Heisenberg W 1981 *Physics and Beyond* (Crow's Nest: Allen & Unwin)

[3] Einsten A, Podolsky B and Rosen N 1935 Can quantum-mechanical description of physical reality be considered complete? *Phys. Rev.* **47** 777

[4] Harris S E, Oshman M K and Byer R L 1967 Observation of tunable optical parametric fluorescence *Phys. Rev. Lett.* **18** 732–4

[5] Bromberg J 1972 Twentieth century—the Born–Einstein letters. Correspondence between Albert Einstein and Max and Hedwig born from 1916 to 1955 with commentaries by Max Born. Trans. by Irene Born. Foreword by Bertrand Russell. Introduction by Werner Heisenberg. London: Macmillan, 1971. pp. xi 240. 3.85 *Br. J. Hist. Sci.* **6** 222–3

[6] Bell J S 1964 On the Einstein Podolsky Rosen paradox *Phys. P. Fiz.* **1** 195

[7] Dehlinger D and Mitchell M W 2002 Entangled photons, nonlocality, and Bell inequalities in the undergraduate laboratory *Am. J. Phys.* **70** 903–10

[8] Bouwmeester D, Pan J-W, Mattle K, Eibl M, Weinfurter H and Zeilinger A 1997 Experimental quantum teleportation *Nature* **390** 575

[9] Poppe A *et al* 2004 Practical quantum key distribution with polarization entangled photons *Opt. Express* **12** 3865–71

[10] Kronowetter F, Würth M, Utschick W, Gross R and Fedorov K G 2024 Imperfect photon detection in quantum illumination *Phys. Rev. Appl.* **21** 014007

[11] Zia D, Dehghan N, D'Errico A, Sciarrino F and Karimi E 2023 Interferometric imaging of amplitude and phase of spatial biphoton states *Nat. Photon.* **17** 1009–16

[12] Bouwmeester D and Zeilinger A 2000 The physics of quantum information: basic concepts *The Physics of Quantum Information* (Berlin: Springer) pp 1–14

[13] Dehlinger D and Mitchell M W 2002 Entangled photon apparatus for the undergraduate laboratory *Am. J. Phys.* **70** 898–902

[14] Beck M 2012 *Quantum Mechanics: Theory and Experiment* (Oxford: Oxford University Press)

[15] Holbrow C H, Galvez E and Parks M E 2002 Photon quantum mechanics and beam splitters *Am. J. Phys.* **70** 260–5

[16] Thorn J J, Neel M S, Donato V W, Bergreen G S, Davies R E and Beck M 2004 Observing the quantum behavior of light in an undergraduate laboratory *Am. J. Phys.* **72** 1210–9

[17] Galvez E J, Holbrow C H, Pysher M J, Martin J W, Courtemanche N, Heilig L and Spencer J 2005 Interference with correlated photons: five quantum mechanics experiments for undergraduates *Am. J. Phys.* **73** 127–40

[18] Carlson J A, Olmstead M D and Beck M 2006 Quantum mysteries tested: an experiment implementing Hardy's test of local realism *Am. J. Phys.* **74** 180–6

[19] Dimitrova T L and Weis A 2010 Single photon quantum erasing: a demonstration experiment *Eur. J. Phys.* **31** 625

[20] Stanley R Q 1996 Question # 45. What (if anything) does the photoelectric effect teach us? *Am. J. Phys.* **64** 839

[21] Milonni P W 1997 Answer to question # 45 ['What (if anything) does the photoelectric effect teach us?', Stanley, R Q Am. J. Phys. 64(7), 839 (1996) *Am. J. Phys.* **65** 11–2

[22] Lamb W E and Scully M O 1969 Polarisation, matière et rayonnement *volume in honour of Alfred Kastler* (Paris: Presses universitaires de France)

[23] Clauser J F 1974 Experimental distinction between the quantum and classical field-theoretic predictions for the photoelectric effect *Phys. Rev.* D **9** 853

[24] Kimble H J, Dagenais M and Mandel L 1977 Photon antibunching in resonance fluorescence *Phys. Rev. Lett.* **39** 691

[25] Burnham D C and Weinberg D L 1970 Observation of simultaneity in parametric production of optical photon pairs *Phys. Rev. Lett.* **25** 84

[26] Grangier P, Roger G and Aspect A 1986 Experimental evidence for a photon anticorrelation effect on a beam splitter: a new light on single-photon interferences *Europhys. Lett.* **1** 173

[27] Schwindt P D D, Kwiat P G and Englert B-G 1999 Quantitative wave-particle duality and nonerasing quantum erasure *Phys. Rev.* A **60** 4285

[28] Schneider M B and LaPuma I A 2002 A simple experiment for discussion of quantum interference and which-way measurement *Am. J. Phys.* **70** 266–71

[29] Gogo A, Snyder W D and Beck M 2005 Comparing quantum and classical correlations in a quantum eraser *Phys. Rev.* A **71** 052103

[30] Eberly J H, Mandel L and Wolf E 2013 *Coherence and Quantum Optics VII: Proc. 7th Rochester Conf. on Coherence and Quantum Optics, held at the (University of Rochester, 7–10 June 1995)* (Berlin: Springer Science & Business Media)

[31] Barrow J D, Davies P C W and Harper C L Jr 2004 Science and Ultimate Reality: Quantum Theory *Cosmology, and Complexity* (Cambridge: Cambridge University Press)

[32] Freedman S J and Clauser J F 1972 Experimental test of local hidden-variable theories *Phys. Rev. Lett.* **28** 938

[33] Clauser J F, Horne M A, Shimony A and Holt R A 1969 Proposed experiment to test local hidden-variable theories *Phys. Rev. Lett.* **23** 880

[34] Aspect A, Grangier P and Roger G 1982 Experimental realization of Einstein-Podolsky-Rosen-Bohm gedanken experiment: a new violation of Bell's inequalities *Phys. Rev. Lett.* **49** 91

[35] Greenberger D M, Horne M A and Zeilinger A 1989 Going beyond Bell's theorem *Bell's Theorem, Quantum Theory and Conceptions of the Universe* (Berlin: Springer) pp 69–72

[36] Pan J-W, Bouwmeester D, Daniell M, Weinfurter H and Zeilinger A 2000 Experimental test of quantum nonlocality in three-photon Greenberger-Horne-Zeilinger entanglement *Nature* **403** 515

[37] Hardy L 1993 Nonlocality for two particles without inequalities for almost all entangled states *Phys. Rev. Lett.* **71** 1665

[38] Torgerson J R, Branning D, Monken C H and Mandel L 1995 Experimental demonstration of the violation of local realism without Bell inequalities *Phys. Lett.* A **204** 323–8

[39] Di Giuseppe G, De Martini F and Boschi D 1997 Experimental test of the violation of local realism in quantum mechanics without Bell inequalities *Phys. Rev.* A **56** 176

[40] White A G, James D F V, Eberhard P H and Kwiat P G 1999 Nonmaximally entangled states: production, characterization, and utilization *Phys. Rev. Lett.* **83** 3103

[41] Brody J and Selton C 2018 Quantum entanglement with Freedman's inequality *Am. J. Phys.* **86** 412–6

[42] Altepeter J B, Jeffrey E R and Kwiat P G 2005 Photonic state tomography *Adv. Atom. Mol. Opt. Phy.* **52** 105–59

[43] Akbar A, Faizan-e-Ilahi and Anwar M S 2021 Quantum process tomography of a magneto-optic transformation *Phys. Lett.* A **406** 127467

[44] Beck M 2007 Comparing measurements of $g^{(2)}(0)$ performed with different coincidence detection techniques *J. Opt. Soc. Am.* B **24** 2972–8

[45] Beck M N and Beck M 2016 Witnessing entanglement in an undergraduate laboratory *Am. J. Phys.* **84** 87–94

[46] Dederick E and Beck M 2014 Exploring entanglement with the help of quantum state measurement *Am. J. Phys.* **82** 962–71

[47] Galvez E J, Malik M and Melius B C 2007 Phase shifting of an interferometer using nonlocal quantum-state correlations *Phys. Rev.* A **75** 020302

[48] Castrillon J, Galvez E J, Rodriguez B A and Calderon-Losada O 2019 A time-energy delayed-choice interference experiment for the undergraduate laboratory *Eur. J. Phys.* **40** 055401

[49] Pysher M J, Galvez E J, Misra K, Wilson K R, Melius B C and Malik M 2005 Nonlocal labeling of paths in a single-photon interferometer *Phys. Rev.* A **72** 052327

[50] Gadway B R, Galvez E J and De Zela F 2008 Bell-inequality violations with single photons entangled in momentum and polarization *J. Phys.* B **42** 015503

[51] Carvioto-Lagos J, Armendariz G, Velázquez V, López-Moreno E, Grether M and Galvez E J 2012 The Hong-Ou-Mandel interferometer in the undergraduate laboratory *Eur. J. Phys.* **33** 1843

[52] Kwiat P G, Waks E, White A G, Appelbaum I and Eberhard P H 1999 Ultrabright source of polarization-entangled photons *Phys. Rev.* A **60** R773

[53] Christensen B G *et al* 2013 Detection-loophole-free test of quantum nonlocality, and applications *Phys. Rev. Lett.* **111** 130406

[54] James D F V, Kwiat P G, Munro W J and White A G 2005 On the measurement of qubits *Asymptotic Theory of Quantum Statistical Inference: Selected Papers* (Singapore: World Scientific) pp 509–38

[55] Bissell L 2011 Experimental realization of efficient, room temperature single-photon sources with definite circular and linear polarizations *PhD Thesis* University of Rochester

IOP Publishing

Quantum Mechanics in the Single-Photon Laboratory (Second Edition)

Muhammad Sabieh Anwar, Faizan-e-Ilahi, Syed Bilal Hyder and Muhammad Hamza Waseem

Chapter 2

Classical nature of light

Before investigating the quantum nature of light, we briefly give an overview of some of its classical properties. This review will serve two purposes. First, it will help familiarize the theoretical and experimental tools required to formally undertake the study of light. Second, it will help us draw and appreciate an analogy between the classical description of coherent light and the quantum description of a beam of single photons. For instance, we will see that polarization analysis for an optical beam fetches us identical results for both quantum and classical descriptions.

Classical optics is based on an elegant synthesis of electricity and magnetism, and sufficiently explains a great number of optical phenomena. Because of the wide scope and range of applicability of classical optics, quantum optics is not often discussed in conventional undergraduate courses on optics, which mostly revolve around only geometrical and physical optics. Some recommended books for classical optics include [1, 2], and [3]. An interesting textbook by Thorne and Blandford [4] covers contemporary topics in optics research.

This chapter describes some possible experimental investigations that form a run-up for the ensuing quantum experiments. These experiments are all performed by the authors at PhysLab, LUMS, and constitute what we call the *Ibn-e-Sahl Corner for Optics*. Students get a chance to perform these experiments as part of their coursework in experimental physics.

2.1 Electromagnetic waves

The theory of light as an electromagnetic wave was developed by James Clerk Maxwell in the nineteenth century and is considered one of the greatest achievements of physics. The study of electricity and magnetism is encapsulated in Maxwell's equations, which can be stated as

$$\nabla \cdot \mathbf{D} = \rho, \tag{2.1}$$

doi:10.1088/978-0-7503-6315-0ch2

$$\nabla \cdot \mathbf{B} = 0, \tag{2.2}$$

$$\nabla \times \mathbf{E} = -\frac{\partial \mathbf{B}}{\partial t}, \text{ and} \tag{2.3}$$

$$\nabla \times \mathbf{H} = \mathbf{J} + \frac{\partial \mathbf{D}}{\partial t}, \tag{2.4}$$

where $\mathbf{E}$ and $\mathbf{B}$ are electric and magnetic fields, ρ is the free charge density, $\mathbf{J}$ is the free current density, $\mathbf{D} = \varepsilon_0 \varepsilon_r \mathbf{E}$ is the electric displacement, and $\mathbf{H}$ is a quantity sometimes called the magnetizing force. The relationship between the magnetic field and magnetizing force is $\mathbf{B} = \mu_0 \mu_r \mathbf{H}$. The variables ε_0, ε_r, μ_0 and μ_r represent the free space permittivity, relative permittivity, free space permeability, and relative permeability, respectively. The relative terms indicate the role of the medium through which the electromagnetic waves propagate and are equal to unity in vacuum.

Equation (2.1) encodes Gauss's law of electrostatics. Equation (2.2) describes the equivalent of Gauss's law for magnetostatics and includes the assumption that free magnetic monopoles do not exist or, at least, they have not been found. Equation (2.3) incorporates Faraday's and Lenz's laws of electromagnetic induction. Equation (2.4) is a statement of Ampere's law wherein the second term on the right-hand side accounts for displacement current.

If there are no free charges or currents, we can combine equations (2.3) and (2.4) to obtain the wave equation

$$\nabla^2 \mathbf{E} = \mu_0 \varepsilon_0 \varepsilon_r \frac{\partial^2 \mathbf{E}}{\partial t^2}, \tag{2.5}$$

which describes electromagnetic waves with the velocity $v = 1/\sqrt{\mu_0 \varepsilon_0 \varepsilon_r}$. In free space ($\varepsilon_r = 1$), this speed is equal to $c \approx 2.998 \times 10^8$ m s^{-1}.

Free space solutions to Maxwell's equations include transverse waves with mutually perpendicular electric and magnetic fields. To illustrate, let's consider a wave propagating in the z-direction and having an electric field along the x-axis. Taking ω to be the angular frequency of the wave, with $E_y = E_z = 0$ and $B_x = B_z = 0$, the Maxwell equations bear electric field solutions of the form

$$E_x(z, t) = E_{x0} \, e^{i(kz - \omega t + \phi)}, \tag{2.6}$$

where E_{x0} represents the amplitude of the electric field, φ denotes the optical phase and $k = 2\pi/\lambda$ is the wave vector. The complex form is often used to simplify mathematical manipulation. Similar solutions exist for the magnetic field and can be expressed as

$$B_y(z, t) = B_{y0} \, e^{i(kz - \omega t + \phi)}. \tag{2.7}$$

Therefore, an electromagnetic wave, such as a light wave, comprises propagating electric and magnetic fields. Waves propagate energy without moving matter.

For any electromagnetic wave, we can compute this energy flow using the Poynting vector,

$$\mathbf{S} = \mathbf{E} \times \mathbf{H}. \tag{2.8}$$

The Poynting vector determines the intensity (which can be described as the energy flow (power) per unit area in W m^{-2}) of an electromagnetic wave. The time-averaged Poynting vector is given by [5]

$$\langle S \rangle = \frac{1}{2} c \varepsilon_0 n E_{x0}^2, \tag{2.9}$$

which shows that the electromagnetic wave intensity is directly proportional to the squared amplitude of the electric field. The quantity $n = \sqrt{\varepsilon_r}$ is called the medium's refractive index.

2.2 Polarization

An electromagnetic wave is transverse and has an electric field $\mathbf{E}$ that is perpendicular to the direction of propagation of the wave. This direction of the electric field is called its *polarization*. If we take an electromagnetic wave propagating in the z-direction in vacuum, the electric field can be pointing at any instant in an arbitrary direction in the xy plane. This is a generalization of the scenario presented in the previous section.

Based on the formulation of equation (2.6), the total electric field can be written in general form as

$$
\begin{aligned}
\mathbf{E} &= E_{x0}\, e^{i(kz-\omega t)}\mathbf{e}_x + E_{y0}\, e^{i(kz-\omega t+\phi)}\mathbf{e}_y \\
&= E_0\, e^{i(kz-\omega t)}\left(\frac{E_{x0}}{E_0}\, \mathbf{e}_x + \frac{E_{y0}}{E_0}e^{i\phi}\, \mathbf{e}_y\right) \\
&= E_0 e^{i(kz-\omega t)}(A\, \mathbf{e}_x + B e^{i\phi}\, \mathbf{e}_y) = E_0\, e^{i(kz-\omega t)}\varepsilon,
\end{aligned}
\tag{2.10}
$$

where $\varepsilon = (A\, \mathbf{e}_x + B e^{i\phi}\, \mathbf{e}_y)$ is a complex unit vector termed as the polarization vector and E_0 is the amplitude of the electrical field, given by $E_0 = \sqrt{E_{x0}^2 + E_{y0}^2}$. The values of the real parameters A, B, and φ determine the polarization state of light. Due to the $e^{-i\omega t}$ factor, the polarization vector gets its time evolution.

2.2.1 The polarization ellipse

We can view the tip of the electric field vector as light advances or propagates. In general, this polarization vector traces an elliptical trajectory (with time), usually termed the polarization ellipse, as shown in figure 2.1, and provides an intuitive understanding of the polarization properties of light. We can express the ellipse using two angular parameters [6], namely the orientation angle denoted as ψ and the ellipticity angle denoted as χ. The former represents the tilt angle, whereas the latter encodes the axial ratio of the ellipse. Let us use the polarization ellipse to look at some special cases of polarization.

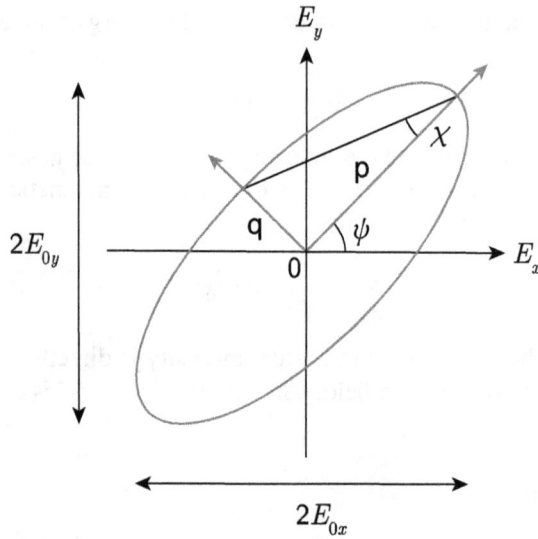

Figure 2.1. For an arbitrary polarization, the polarization ellipse can be represented in terms of the orientation angle ψ and the ellipticity angle χ. Here, p and q denote the semi-major and semi-minor axes, respectively.

Light is linearly polarized if the polarization vector does not change direction as light propagates. In this case, the polarization vector has a zero phase shift ($\phi = 0$) between the x and y components. Hence, the polarization vector is real and given by the real coefficients A and B,

$$\varepsilon = A\,\mathbf{e}_x + B\,\mathbf{e}_y. \tag{2.11}$$

Some common examples of linear polarization include horizontal (where $A = 1$ and $B = 0$), vertical (where $A = 0$ and $B = 1$), diagonal (where $A = B = 1/\sqrt{2}$), and anti-diagonal (where $A = 1/\sqrt{2}$ and $B = -1/\sqrt{2}$).

If the polarization vector rotates to map out a circle as the wave propagates, the light is called circularly polarized. Circularly polarized light comes in two flavors depending on the clockwise or anti-clockwise rotation of the electric field vector. The polarization vector is composed of equal parts of x and y components with a phase difference of $\phi = \pm\pi/2$ between them, resulting in the aforementioned two options for the polarization vector

$$\varepsilon = \frac{1}{\sqrt{2}}(\mathbf{e}_x \pm i\,\mathbf{e}_y). \tag{2.12}$$

These linear and circular polarization states are called degenerate polarization states [6] and are illustrated in figure 2.2, which shows the locus of the tip of the electric field vector.

For an arbitrary polarization state, the polarization vector traces an ellipse. In this case, either the amplitudes of the x and y components are different, or the phase difference between the two components is neither $0°$ nor $90°$.

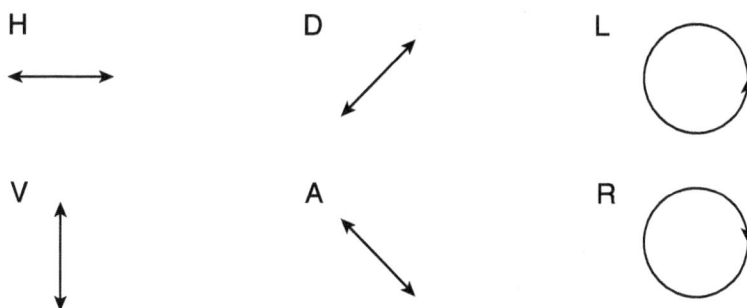

Figure 2.2. Convenient representations of the polarization vector showing degenerate polarization states: horizontal (H), vertical (V), diagonal (D), anti-diagonal (A), right circular (R), and left circular (L). The lines or circles represent the locus of the tip of the electric field vector viewed in the direction of propagation as the wave progresses.

Light is considered unpolarized if there are random fluctuations in the polarization vector with time. The exact parameters cannot be determined in this case. However, statistical descriptions can still be made using Stokes parameters, which are discussed in a later section.

2.2.2 Manipulating polarization

We have mentioned commonly studied polarized states. In optical experiments, we frequently need to modify polarization. For this purpose, we use several kinds of optical elements. Let us briefly discuss the functional description of a few of these elements.

A *linear polarizer* transmits only that component of the polarization vector ε which is parallel to its preferred transmission axis $\mathbf{e}_\theta$. The transmitted field is given by

$$\mathbf{E}_t = E_0 \, e^{i(kz-\omega t)}(\varepsilon \cdot \mathbf{e}_\theta)\hat{e}_\theta. \tag{2.13}$$

It is evident that the transmitted field is linearly polarized along $\hat{e}_\theta$ and its amplitude is downscaled by $|\varepsilon \cdot \mathbf{e}_\theta|$.

Many optical elements are based on birefringent crystals. Birefringent materials exhibit two different refractive indices depending on the polarization and direction of propagation of light that enters them. These crystals can change the polarization of light passing through them and can also separate an arbitrarily polarized optical ray into two rays with orthogonal polarizations. These emergent rays are termed the ordinary ray and the extraordinary ray. They experience different refractive indices labeled as n_o and n_e, respectively. The refractive index controls the phase velocity of waves. If these indices are different, the different components of light accrue a phase difference φ as they propagate through the crystal as shown in figure 2.3, resulting in a modified polarization state.

A *wave plate* precisely achieves this effect, modifying the relative phase shift between the orthogonal components of the polarization vector. Quarter-wave plates (QWPs) that induce a phase difference of $\lambda/4$, and half-wave plates (HWPs) that

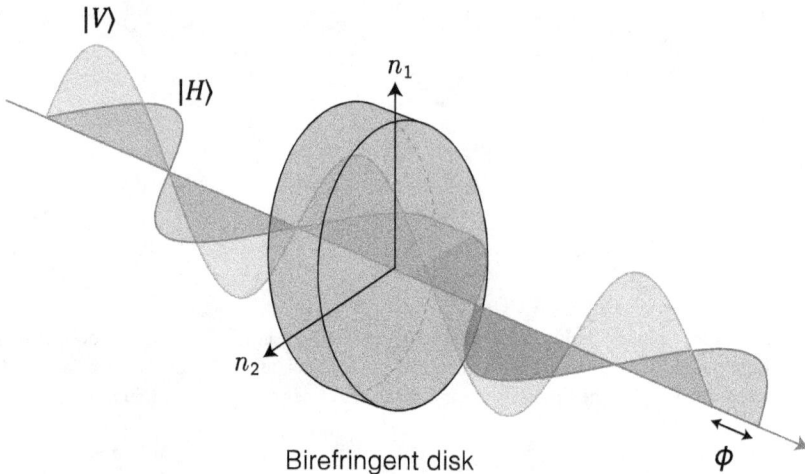

Figure 2.3. A birefringent material exhibits different refractive index for different components of light and induces a phase between them. The H component goes through the fast axis while the V component goes through the slow axis, resulting in a phase φ between the two components.

induce a phase difference of $\lambda/2$ are the most commonly used wave plates. Depending on its orientation and the input polarization, a QWP can turn a linear polarization into an elliptical (circular) polarization, or vice versa. Similarly, a HWP can rotate a linear polarization, and the orientation of the HWP controls the rotation with respect to the input polarization.

A polarizing beam splitter (PBS) resolves the incident light into its orthogonal polarization components [7]. In the kind of PBS that we will use, the horizontal and vertical components emerge at right angles.

Using a laser and the polarization-modifying components that we have just described, we can perform many experiments focused on generating arbitrary polarization states, manipulating them, and detecting the resultant light beams. In order to mathematically analyze the results and predictions of these experiments, Jones calculus proves to be a valuable tool and is now described.

2.2.3 Jones calculus

Jones calculus maps any pure polarization state to a 2×1 column vector and maps polarization manipulating elements to 2×2 matrices. Consider again the arbitrary polarization vector

$$\varepsilon = A\, \mathbf{e}_x + B e^{i\phi}\, \mathbf{e}_y. \tag{2.14}$$

Since we have a normalized polarization vector, the real parameters A and B satisfy $A^2 + B^2 = 1$, and φ is the phase difference between the orthogonal components. Conventionally, the horizontal and vertical components are taken to be $\mathbf{e}_H = \mathbf{e}_x$ and $\mathbf{e}_V = \mathbf{e}_y$. These vectors are then used as basis vectors

Table 2.1. Jones vectors corresponding to canonical polarization states. The angles α, 45° and −45° are with respect to the horizontal axis of the lab frame of reference.

Polarization	Symbol	Jones vector
Linear at angle α	ε_α	$\begin{pmatrix} \cos\alpha \\ \sin\alpha \end{pmatrix}$
Horizontal (linear along x-axis)	ε_H	$\begin{pmatrix} 1 \\ 0 \end{pmatrix}$
Vertical (linear along y-axis)	ε_V	$\begin{pmatrix} 0 \\ 1 \end{pmatrix}$
Diagonal (linear at 45°)	ε_D	$\frac{1}{\sqrt{2}}\begin{pmatrix} 1 \\ 1 \end{pmatrix}$
Anti-diagonal (linear at −45°)	ε_A	$\frac{1}{\sqrt{2}}\begin{pmatrix} 1 \\ -1 \end{pmatrix}$
Left circular	ε_L	$\frac{1}{\sqrt{2}}\begin{pmatrix} 1 \\ i \end{pmatrix}$
Right circular	ε_R	$\frac{1}{\sqrt{2}}\begin{pmatrix} 1 \\ -i \end{pmatrix}$
Arbitrary	ε	$\begin{pmatrix} A \\ B\,e^{i\phi} \end{pmatrix}$

$$\mathbf{e}_H \equiv \begin{pmatrix} 1 \\ 0 \end{pmatrix} \text{ and } \mathbf{e}_V \equiv \begin{pmatrix} 0 \\ 1 \end{pmatrix} \tag{2.15}$$

to describe any arbitrary polarization vector as follows:

$$\varepsilon = A\,\mathbf{e}_H + Be^{i\phi}\,\mathbf{e}_V = \begin{pmatrix} A \\ B\,e^{i\phi} \end{pmatrix}. \tag{2.16}$$

We call $\mathbf{e}_H$, $\mathbf{e}_V$ the *canonical basis* and the representation of any polarization state as a column vector in this basis is called a *Jones vector*. The Jones vectors for degenerate and elliptical polarization states are listed in table 2.1.

Earlier, we described certain optical elements that can manipulate the polarization of a wave. The mathematical counterparts of these objects are matrices which change one Jones vector into another. These polarization modifiers are called *Jones matrices*. The Jones matrices for the commonly used optical experiments are listed in table 2.2.

To compute the effect of a polarization-changing element on a particular polarization, the Jones matrix of the element is simply left-multiplied to the Jones vector corresponding to the original polarization state. Usually, the polarization vector is normalized before performing Jones calculus manipulations. The output of a particular polarization element is then a unit vector multiplied by a complex constant, whose phase and amplitude determine the phase and amplitude of the electric field. The unit vector itself encapsulates the *relative* sizes of the two basis terms, and the *relative* phase determines the polarization state, which in turn can be represented by the polarization ellipse described earlier.

Table 2.2. Jones matrices corresponding to commonly used polarization-changing optical elements. The axis orientations, represented by θ, are with respect to the horizontal axis of the lab frame.

Optical element	Symbol	Jones matrix
Linear polarizer (axis at θ)	$\mathbf{J}_P(\theta)$	$\begin{pmatrix} \cos^2\theta & \sin\theta\cos\theta \\ \sin\theta\cos\theta & \sin^2\theta \end{pmatrix}$
HWP (fast axis at θ)	$\mathbf{J}_{HWP}(\theta)$	$\begin{pmatrix} \cos 2\theta & \sin 2\theta \\ \sin 2\theta & -\cos 2\theta \end{pmatrix}$
QWP (fast axis at θ)	$\mathbf{J}_{QWP}(\theta)$	$\begin{pmatrix} \cos^2\theta + i\sin^2\theta & (1-i)\sin\theta\cos\theta \\ (1-i)\sin\theta\cos\theta & \sin^2\theta + i\cos^2\theta \end{pmatrix}$

If a beam with a certain polarization state ε_0 passes through a series of polarization-modifying elements with Jones matrices $\mathbf{J}_1, \mathbf{J}_2, \cdots, \mathbf{J}_n$, then the resultant polarization ε_n is given by

$$\varepsilon_n = \mathbf{J}_n \cdots \mathbf{J}_2\mathbf{J}_1\varepsilon_0 = \mathbf{J}_{\text{eff}}\varepsilon_0, \tag{2.17}$$

i.e., the Jones matrices representing the respective optical elements are ordered from right to left. The output polarization is generally dependent on the order of the polarization manipulators. Moreover, any series of optical elements maps to an effective Jones matrix $\mathbf{J}_{\text{eff}}$, which can be found by multiplying the respective Jones matrices taken in the correct order.

Jones calculus works perfectly for polarized light but does not apply to partially polarized or unpolarized light. In these cases, we use Stokes parameters, which in contrast to the *amplitude* description in the Jones calculus are based on the *intensity* representation of polarization states and can even describe unpolarized light.

2.2.4 Stokes parameters

Stokes parameters describe the polarization of light in terms of four observables. There are several ways to express the Stokes parameters, one of which is [6]

$$\begin{pmatrix} S_0 \\ S_1 \\ S_2 \\ S_3 \end{pmatrix} = \begin{pmatrix} I_H + I_V \\ I_D - I_A \\ I_L - I_R \\ I_H - I_V \end{pmatrix} = \begin{pmatrix} E_{0x}^2 + E_{0y}^2 \\ 2E_{0x}E_{0y}\cos\phi \\ 2E_{0x}E_{0y}\sin\phi \\ E_{0x}^2 - E_{0y}^2 \end{pmatrix}. \tag{2.18}$$

Another equivalent expression of the Stokes parameters [8] makes apparent their relation to the polarization ellipse parameters ψ and χ, as follows:

$$\mathbf{S} = \begin{pmatrix} S_1 \\ S_2 \\ S_3 \end{pmatrix} = \begin{pmatrix} \cos(2\chi)\sin(2\psi) \\ \sin(2\chi) \\ \cos(2\chi)\cos(2\psi) \end{pmatrix} \tag{2.19}$$

where $S_1^2 + S_2^2 + S_3^2 = S_0^2$, $0 \leqslant \psi \leqslant \pi$ and $-\pi/4 < \chi \leqslant \pi/4$. For completely polarized light, S_0 is 1; for partially polarized light, it is less than 1; for completely unpolarized light, it is 0. The Stokes parameters can also be graphically represented on the Bloch sphere, shown in figure 2.4(a). Here, S_1, S_2, and S_3 represent the three-dimensional coordinate axes, and $\mathbf{S}$ vector represents the particular polarization state. Table 2.3 lists the Stokes parameters for canonical polarization states.

Another graphical representation called the Poincaré sphere, shown in figure 2.4(b), is obtained if the Stokes parameters are defined as

$$\begin{pmatrix} S_0 \\ S_1 \\ S_2 \\ S_3 \end{pmatrix} = \begin{pmatrix} I_H + I_V \\ I_H - I_V \\ I_D - I_A \\ I_L - I_R \end{pmatrix}. \tag{2.20}$$

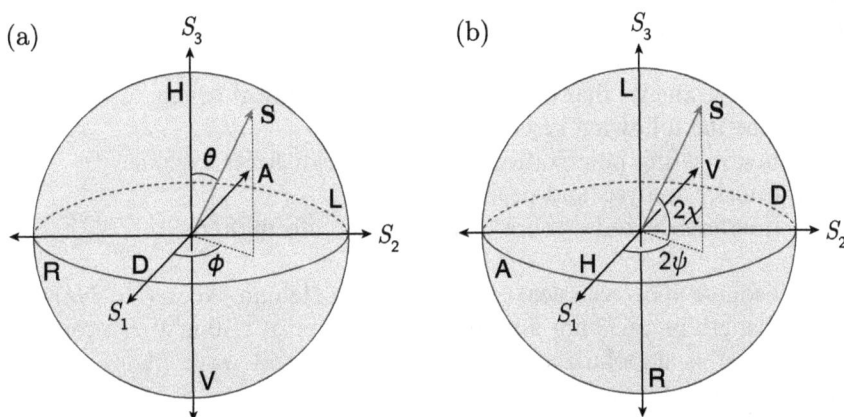

Figure 2.4. (a) The Bloch sphere representation of polarization, where the Stokes parameters S_1, S_2, and S_3 represent the three-dimensional coordinate axes and $\mathbf{S}$ vector represents the particular polarization state. (b) A sister representation to the Bloch sphere is the Poincaré sphere, which maps the degenerate polarization states to different Stokes axes. The sphere shows that the Stokes parameters are related to the polarization ellipse angles ψ and χ shown in figure 2.1.

Table 2.3. Stokes parameters for degenerate polarization states, which are perfectly polarized. We use the definitions in equation (2.18).

Polarization	S_0	S_1	S_2	S_3
Horizontal (H)	1	0	0	1
Vertical (V)	1	0	0	−1
Diagonal (D)	1	1	0	0
Anti-diagonal (A)	1	−1	0	0
Left circular (L)	1	0	1	0
Right circular (R)	1	0	−1	0

Comparing these definitions with those in equation (2.18), it is evident that the Poincaré and Bloch spheres are similar but map the degenerate polarization states to different Stokes axes. Although the Poincaré representation is ubiquitous in classical optics, we prefer the Bloch picture because it conforms to commonplace quantum optical description of polarization, which is the main thrust of this book.

Similar to the Jones calculus, the effect of polarization-changing elements on the polarization state can be computed by multiplying 4×4 Mueller matrices with the vector of equation (2.18), a system termed Mueller calculus. On the Bloch sphere, these polarization manipulations are mapped as rotations and scaling of the **S** vector.

2.3 Preparatory experimental explorations

Before we embark on our expedition of setting up experiments that demonstrate quantum effects, it is instructive to describe a few experiments within the realm of classical optics. These experiments will brush up many mathematical and physical concepts in optics, and lay the groundwork for several useful experimental techniques and strategies that are employed on the optical bench.

We describe the following key experiments:
- C1: Investigating polarization of light through Jones calculus
- C2: Fourier analysis and peanut plots
- C3: Interference and erasure of which-way information

The light source in our classical experiments is a Helium–Neon (He–Ne) gas laser with a wavelength of ≈ 633 nm and an optical power of $\leqslant 10$ mW at room temperature. The laser is unpolarized. Hence, polarizers and wave plates are used to generate the required polarization states.

For light detection, we use silicon photodiodes, which output a current proportional to the intensity of the light beam incident on them. We convert the current to a proportional voltage through a current-to-voltage (IV) converter, which comprises a transimpedance amplifier. With a photodiode, instead of measuring the electric field associated with the optical wave, we rather measure the *square* of the electric field, which is proportional to the optical intensity I (measured in $W \, m^{-2}$) impinging on it. This is also in line with the average of the Poynting vector (equation (2.9)). Therefore, we define the optical intensity as the squared magnitude of the polarization vector

$$,I \equiv \varepsilon^* \cdot \varepsilon = |\varepsilon|^2. \tag{2.21}$$

2.4 C1: Investigating polarization of light through Jones calculus

Referring to the schematic in figure 2.5a, we generate different linear polarization states through a polarizer and an HWP. Light from a Helium–Neon (He–Ne) laser passes through a linear polarizer oriented at α with respect to the horizontal and

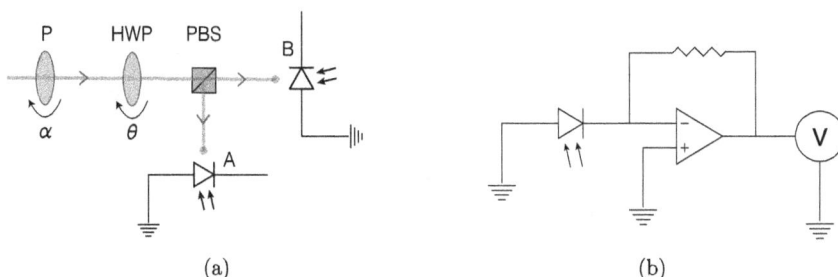

(a) (b)

Figure 2.5. (a) Schematic diagram of an experiment to study the polarization of light and practicing Jones calculus. Light from a 633 nm laser passes through a polarizer (P) oriented at α, a HWP oriented at θ and falls on a PBS, which lets vertically polarized light transmit and fall on photodetector A and horizontally polarized reflect toward photodetector B. (b) An IV converter converts the photodetector's output current into a proportional voltage readable by a voltmeter.

then an HWP oriented at θ with respect to the horizontal. The polarization of light coming out of the HWP can be computed using Jones calculus and is

$$
\begin{aligned}
\varepsilon &= \mathbf{J}_{HWP}(\theta)\varepsilon_\alpha \\
&= \begin{pmatrix} \cos 2\theta & \sin 2\theta \\ \sin 2\theta & -\cos 2\theta \end{pmatrix}\begin{pmatrix} \cos \alpha \\ \sin \alpha \end{pmatrix} \\
&= \begin{pmatrix} \cos(2\theta - \alpha) \\ \sin(2\theta - \alpha) \end{pmatrix}.
\end{aligned}
\tag{2.22}
$$

Subsequently, the light is incident on a PBS, which separates the two orthogonal components of the field, which are then detected as intensities impingent on the two photodiodes A and B. These intensities are given by $I_A = \cos^2(2\theta - \alpha)$ and $I_B = \sin^2(2\theta - \alpha)$, respectively. The simulated and experimentally measured intensities are plotted in figure 2.6.

Each photodiode is connected to an IV converter (schematically shown in figure 2.5b), which converts the photodiode output current, which is proportional to light intensity, into a voltage which is then measured through a voltmeter. The measured intensities plotted in figure 2.6 show excellent agreement with curve fits of intensity expressions obtained from Jones calculus. As shown in the simulated curves, the two intensities are orthogonal to each other, confirming the basic function of the PBS. Moreover, turning the HWP shows a periodicity of 90°, while turning the polarizer shows a periodicity of 180°. Both of these features are corroborated in the simulated and experimentally achieved results.

2.5 C2: Fourier analysis and peanut plots

We can perform another interesting experiment to analyze the polarization state of light. This method is based on Fourier analysis and involves the Stokes parametric description of polarization. As shown in figure 2.7, a polarizer oriented at β and a QWP oriented at θ are used to generate an arbitrarily polarized laser beam. Another

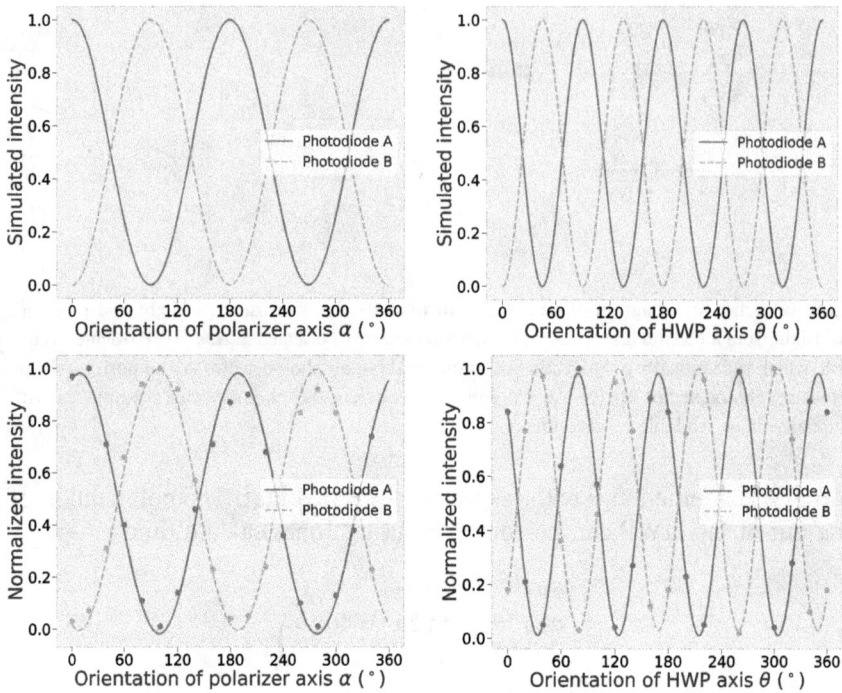

Figure 2.6. Simulated and experimental results of the experiment of figure 2.5. Blue plots represent detector A, while red plots represent detector B. All intensities are normalized. (a) Simulated variation in intensity with respect to polarizer orientation α when the HWP is oriented at $0°$. (b) Simulated variation in intensity with respect to HWP orientation θ when the polarizer is oriented at $0°$. (c) Measured variation in intensity with respect to polarizer orientation. (d) Measured variation in intensity with respect to HWP orientation. The solid squares represent the measurements, while the smooth plots represent the curve fits.

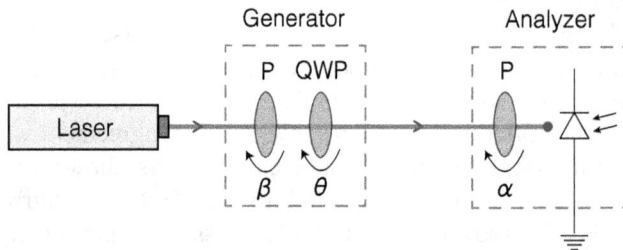

Figure 2.7. Schematic diagram of an experiment to study the polarization of light via Fourier analysis and Stokes parameters. Light from a 633 nm laser passes through a polarizer (P) oriented at β, a QWP oriented at θ, another polarizer oriented at α and finally falls on a photodetector. The two dotted boxes represent the polarization state generator and the polarization state analyzer, in order from left to right.

polarizer oriented at α is then used as an analyzer. All angular orientations are quoted with respect to the horizontal axis of the lab frame. The electric field prior to hitting the photodetector can be calculated in the usual fashion through Jones calculus:

$$\varepsilon = \mathbf{J}_P(\alpha)\mathbf{J}_{QWP}(\theta)\varepsilon_\beta$$

$$= \begin{pmatrix} \cos^2\alpha & \sin\alpha\cos\alpha \\ \sin\alpha\cos\alpha & \sin^2\alpha \end{pmatrix}\begin{pmatrix} \cos^2\theta + i\sin^2\theta & (1-i)\sin\theta\cos\theta \\ (1-i)\sin\theta\cos\theta & \sin^2\theta + i\cos^2\theta \end{pmatrix}\begin{pmatrix} \cos\beta \\ \sin\beta \end{pmatrix}$$

$$= \begin{pmatrix} \dfrac{1+i}{2}\cos\alpha[\cos(\alpha-\beta) - i\cos(\alpha+\beta-2\theta)] \\ \dfrac{1-i}{2}\sin\alpha[i\cos(\alpha-\beta) + \cos(\alpha+\beta-2\theta)] \end{pmatrix}. \tag{2.23}$$

Finally, the intensity is calculated as

$$I = \varepsilon^*\!\cdot\!\varepsilon = \frac{1}{4}(2 + \cos(2(\alpha-\beta)) + \cos(2(\alpha+\beta-2\theta)))$$

$$= \frac{1}{2} + \frac{1}{2}\cos(2(\theta-\beta))\cos(2\theta)\cos(2\alpha) + \frac{1}{2}\cos(2(\theta-\beta))\sin(2\theta)\sin(2\alpha), \tag{2.24}$$

which is a Fourier series in α of the form [9]

$$I = \frac{S_0}{2} + \frac{S_3}{2}\cos(2\alpha) + \frac{S_1}{2}\sin(2\alpha). \tag{2.25}$$

We can show that the Fourier coefficients S_0, S_3, and S_1 in equation (2.25) are, in fact, the respective Stokes parameters of the light leaving the QWP. Consider the polarization generated by the polarization generator

$$\mathbf{J}_{QWP}(\theta)\varepsilon_\beta = \begin{pmatrix} \cos^2\theta + i\sin^2\theta & (1-i)\sin\theta\cos\theta \\ (1-i)\sin\theta\cos\theta & \sin^2\theta + i\cos^2\theta \end{pmatrix}\begin{pmatrix} \cos\beta \\ \sin\beta \end{pmatrix}$$

$$= \begin{pmatrix} \dfrac{1+i}{2}(\cos\beta - i\cos(\beta-2\theta)) \\ \dfrac{1+i}{2}(\sin\beta + i\sin(\beta-2\theta)) \end{pmatrix}. \tag{2.26}$$

Then, S_0, S_3, and S_1 can be calculated from the components of this vector:

$$S_0 = I_H + I_V = E_{0x}^2 + E_{0y}^2$$

$$= \left|\frac{1+i}{2}(\cos\beta - i\cos(\beta-2\theta))\right|^2 + \left|\frac{1+i}{2}(\sin\beta + i\sin(\beta-2\theta))\right|^2 \tag{2.27}$$

$$= 1,$$

$$S_3 = I_H - I_V = E_{0x}^2 - E_{0y}^2$$

$$= \left|\frac{1+i}{2}(\cos\beta - i\cos(\beta-2\theta))\right|^2 - \left|\frac{1+i}{2}(\sin\beta + i\sin(\beta-2\theta))\right|^2 \tag{2.28}$$

$$= \cos(2(\theta-\beta))\cos(2\theta), \text{ and}$$

$$S_1 = I_D - I_A = \left| E_D \right|^2 - \left| E_A \right|^2 = \left| \frac{E_{0x} + E_{0y}}{\sqrt{2}} \right|^2 - \left| \frac{E_{0x} - E_{0y}}{\sqrt{2}} \right|^2 \quad (2.29)$$
$$= \cos(2(\theta - \beta))\sin(2\theta),$$

which are consistent with equations (2.24) and (2.25) (and, comparing with equation (2.19), also imply that $\theta = \psi$ and $\theta - \beta = \chi$). Therefore, if we have intensity measurements at a number of analyzer orientations for a given polarization state, we can make a Fourier curve fit to find the three Stokes parameters. This method, however, does not fetch us S_2, which provides information about the handedness of the polarization state. Hence, the analyzer cannot differentiate between left circularly polarized and right circularly polarized light. To determine S_2, at least one more QWP needs to be added to the analyzer.

In our warm-up experiments, we use the polarization state generator described above to generate a number of polarization states. For each state, the analyzing polarizer is oriented in steps of 10° from 0° through 360° and the photodiode's output is measured. The aforementioned technique is then used to determine the respective Stokes parameters, and the experimental results are summarized in table 2.4. See the column titled 'Fourier fit'. The measured results show good agreement with theoretical predictions.

We now describe an alternative technique [10] to compute the Stokes parameters in the same experiment. This technique is adapted from antenna polarimetry and will be revisited when we discuss the quantum experiments. Suppose that we obtained the intensity profile for an arbitrary polarization with respect to analyzer orientations as described above. Now, if we make a plot between $\sqrt{I(\alpha)} \cos \alpha$ and $\sqrt{I(\alpha)} \sin \alpha$, we obtain a peanut-shaped curve, which geometers call a 'hippopede' and is shown in figure 2.8.

It is evident from the figure that the polarization ellipse and the hippopede have the same orientation ψ and ellipticity χ angles. Therefore, ψ can be directly measured from the tilt of the hippopede, while χ can be determined by measuring the axial ratio (q/p) and subsequently using the formula $\chi = \pm \tan^{-1}(q/p)$. The Stokes parameters can then be computed using equation (2.19). Note the sign ambiguity of χ and resultantly of S_2, which implies that once again, this experimental scheme does not tell us about the polarization handedness. Table 2.4 enlists the Stokes parameters for degenerate polarization states obtained using this technique. See the column titled 'Hippopede method'. The results show good agreement with both Fourier curve fit results, as well as theoretical predictions. We will revisit these peanut-shaped patterns for quantum and classical light in the experimental results in chapter 5.

2.6 C3: Interference and erasure of which-way information

In order to see the interference of light, we employ a Mach–Zehnder interferometer [11], shown in figure 2.9. This apparatus divides the input light beam into two beams and then recombines them. The path difference between the two component beams

Table 2.4. Results of optical polarimetry of degenerate polarization states, comparing theoretical predictions and experimental outcomes. Our technique does not allow us to measure the parameter S_2. N.A. stands for 'not available'.

Polarization state	Fourier fit	Hippopede method	Prediction
Horizontal (H)	$S_0 = 1.00 \pm 0.02$	1.00 ± 0.00	1
	$S_3 = 0.97 \pm 0.02$	1.00 ± 0.00	1
	$S_1 = 0.00 \pm 0.02$	0.00 ± 0.02	0
Vertical (V)	$S_0 = 1.00 \pm 0.02$	1.00 ± 0.00	1
	$S_3 = -0.97 \pm 0.02$	-0.99 ± 0.00	-1
	$S_1 = 0.10 \pm 0.02$	0.07 ± 0.02	0
Diagonal (D)	$S_0 = 1.00 \pm 0.04$	1.00 ± 0.00	1
	$S_3 = 0.04 \pm 0.04$	0.04 ± 0.02	0
	$S_1 = 0.94 \pm 0.04$	0.99 ± 0.00	1
Anti-diagonal (A)	$S_0 = 1.00 \pm 0.02$	1.00 ± 0.00	1
	$S_3 = 0.00 \pm 0.02$	0.00 ± 0.02	0
	$S_1 = -0.98 \pm 0.02$	-1.00 ± 0.00	-1
Left circular (L)	$S_0 = 1.00 \pm 0.01$	1.00 ± 0.00	1
	$S_3 = 0.07 \pm 0.01$	0.00 ± 0.00	0
	$S_1 = -0.01 \pm 0.01$	0.05 ± 0.02	0
	$S_2 = $ N.A.	N.A.	1
Right circular (R)	$S_0 = 1.00 \pm 0.02$	1.00 ± 0.00	1
	$S_3 = -0.02 \pm 0.02$	0.00 ± 0.00	0
	$S_1 = 0.04 \pm 0.02$	0.02 ± 0.02	0
	$S_2 = $ N.A.	N.A.	-1

results in an interference pattern. Therefore, the He–Ne laser beam is fed into a beam splitter (BS), which lets half the incoming light pass through it and reflects the rest. Each of the two beams is then reflected off a mirror (M) in its respective path. The beams combine at the second BS and fall on a white screen where interference fringes can be seen. We note that each BS is a non-polarizing beam splitter. If one of the two interferometer beams is blocked, no interference will be apparent.

A Mach–Zehnder interferometer is also a useful device to study not only the basic principles of polarization but also the erasure of which-way information. This erasure experiment underlies an important discussion in quantum mechanics and harbors a profound foundational significance. Let's see how.

Consider the interferometric setup in figure 2.10 where we have added three linear polarizers: one before the light entering the interferometer and one in each of the two interferometer paths. We set the polarizer P_0 at 45° and P_1 at 0°, and record the interference pattern for two orientations of polarizer P_2: at 0° and at 90°. The outcomes achieved are shown in figure 2.11, indicating that we observe no interference when the axes of the polarizers P_1 and P_2 are perpendicular to each

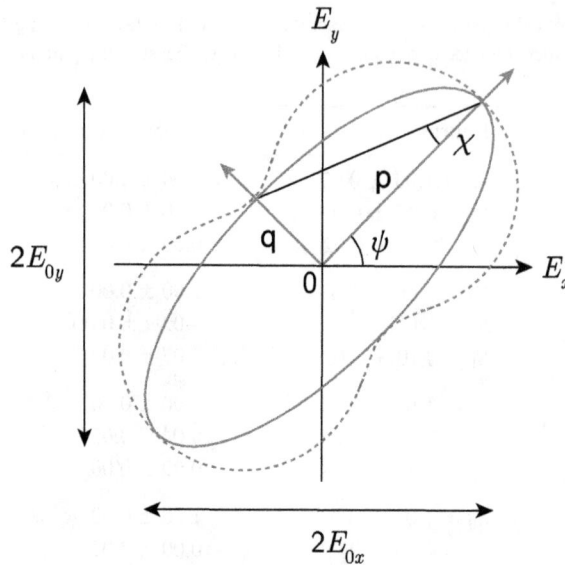

Figure 2.8. Relation of the polarization ellipse (blue solid) and the hippopede (red dashed) for an arbitrary polarization. Angular parameters of the ellipse include the orientation angle ψ and the ellipticity angle χ. Furthermore, p and q denote the semi-major and semi-minor axes, respectively. Adapted with permission from [10]. © The Optical Society.

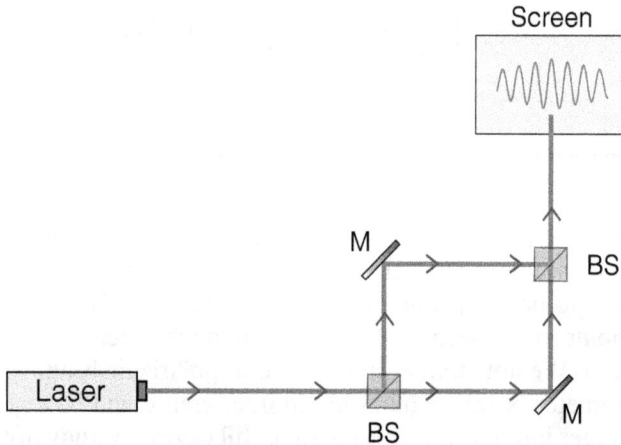

Figure 2.9. Schematic diagram of a Mach–Zehnder interferometer. Light from a 633 nm laser enters an interferometer comprising two BSs and two mirrors (M). Light from the two paths of the interferometer combine and fall on a screen.

other, implying that light has taken only one of the two paths. On the contrary, high visibility fringes are observed when the axes are mutually parallel. In such a case, it is impossible to determine the path taken by the light as it leaves the interferometer. From another perspective, the presence of an interference pattern shows that the light has taken both paths simultaneously.

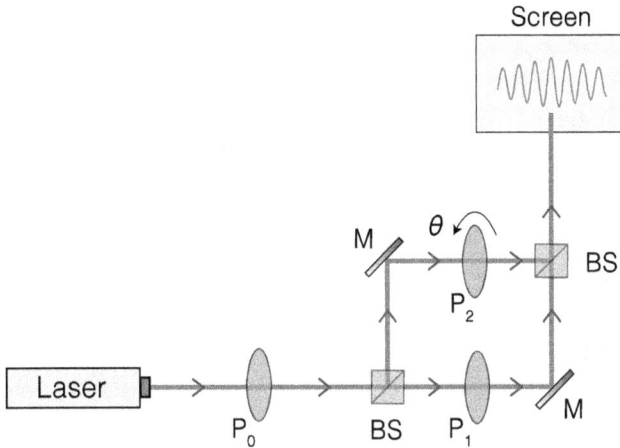

Figure 2.10. Schematic diagram of the Mach–Zehnder interferometer with three polarizers P_0, P_1, and P_2.

(a)   (b)

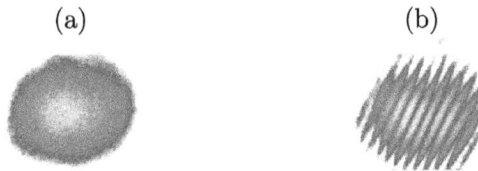

Figure 2.11. Referring to the experiment in figure 2.10, with polarizer P_0 set at 45°, (a) when polarizers P_1 and P_2 are crossed (oriented orthogonally), no interference is visible; (b) but when polarizers P_1 and P_2 are oriented with parallel axes, interference fringes are observed.

Another way to state this observation is in terms of information. In the absence of which-way information, light takes both the available paths in the interferometer, and hence interference fringes are observed. On the other hand, we extract which-way information by using crossed polarizers, which effectively tag the path of light. If we had desired, we would measure the polarization of photon emerging from the second beam splitter BS and the polarization would immediately tell us the path. In this case, no interference is observed, implying that light takes only one of the two paths. And we *do know that* path! Which-way information and interference are mutually exclusive. A knowledge of the path precludes interference.

We can study the experiment using Jones calculus [12]. The polarization state of light entering the interferometer after P_0 can be expressed as

$$\varepsilon_{\text{in}} = \frac{1}{\sqrt{2}} \begin{pmatrix} 1 \\ 1 \end{pmatrix}, \tag{2.30}$$

while the polarization state of light leaving the interferometer and hitting the screen can be computed as

$$\varepsilon_{\text{out}} = r\mathbf{J}_{P2}r_{\text{m}}r\varepsilon_{\text{in}} + e^{i\phi}tr_{\text{m}}\mathbf{J}_{P1}t\varepsilon_{\text{in}}, \tag{2.31}$$

where r and t are the reflection and transmission coefficients of the BS, r_m represents the reflection coefficient of each mirror, and $e^{i\phi}$ represents the phase difference between the two beams due to a small path difference within the coherence length[1] of the laser. We take the mirror reflection coefficients r_m to be 1, and for a 50:50 BS, the coefficients r and t are each equal to 1/2. Taking P_1 to be oriented horizontally and P_2 to be oriented at θ, the output field becomes

$$
\begin{aligned}
\varepsilon_{\text{out}} &= \frac{1}{2}\begin{pmatrix} \cos^2\theta & \sin\theta\cos\theta \\ \sin\theta\cos\theta & \sin^2\theta \end{pmatrix}\frac{1}{2}\frac{1}{\sqrt{2}}\begin{pmatrix} 1 \\ 1 \end{pmatrix} + e^{i\phi}\frac{1}{2}\begin{pmatrix} 1 & 0 \\ 0 & 0 \end{pmatrix}\frac{1}{2}\frac{1}{\sqrt{2}}\begin{pmatrix} 1 \\ 1 \end{pmatrix} \\
&= \frac{1}{4\sqrt{2}}\begin{pmatrix} \cos^2\theta + \cos\theta\sin\theta + e^{i\phi} \\ \cos\theta\sin\theta + \sin^2\theta \end{pmatrix},
\end{aligned}
\tag{2.32}
$$

for which the detectable intensity is computed to be

$$
\begin{aligned}
I &= \varepsilon_{\text{out}}^* \cdot \varepsilon_{\text{out}} \\
&= \frac{1}{32}(2 + \sin(2\theta) + \sin(2\theta)\cos(\phi) + 2\cos^2(\theta)\cos(\phi)).
\end{aligned}
\tag{2.33}
$$

It can be seen that for P_2 set at $\theta = 0°$, I becomes $(1 + \cos\phi)/16$, which describes the dependence on φ of the presence of interference fringes. On the other hand, if P_2 is set at $\theta = 90°$, I becomes 1/16, giving a constant intensity pattern with no dependence on the path difference φ and hence no interference. Figure 2.12 shows observed intensity profiles for a number of different orientations θ of P_2. Our observations are in agreement with the mathematical analysis.

Next, let us consider an interesting extension of the interferometer, as shown in figure 2.13. We place another polarizer P_3 *after* the interferometer, and setting the polarizers P_0 at 45°, P_1 at 0° and P_2 at 90°, observe the interference pattern for different orientations ψ of P_3. Mathematically, this means putting $\theta = 90°$ in

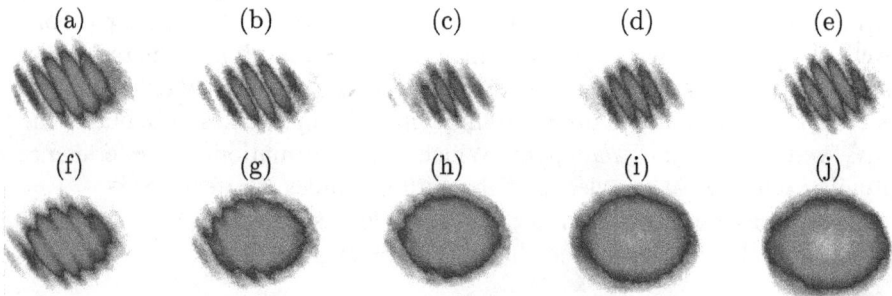

Figure 2.12. Observed intensity profiles for the experiment depicted in figure 2.10. P_0 and P_1 are oriented at 45° and 0°, respectively. P_2 is oriented at (a) 0°, (b) 10°, (c) 20°, (d) 30°, (e) 40°, (f) 50°, (g) 60°, (h) 70°, (i) 80°, and (j) 90°.

---

[1] Coherence length is the propagation distance over which a wave can retain its ability to interfere.

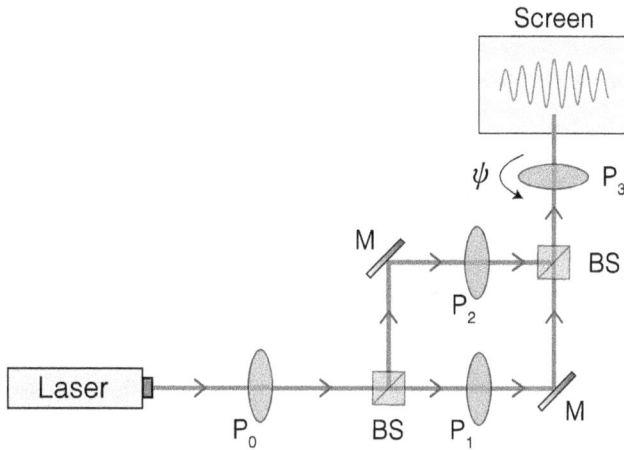

Figure 2.13. Schematic diagram of the extended Mach–Zehnder interferometer with four polarizers P_0, P_1, P_2, and P_3.

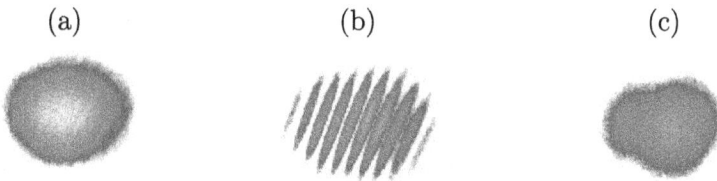

Figure 2.14. Referring to the experiment in figure 2.13, with polarizer P_0 set at 45°, P_1 set at 0°, and P_2 set at 90° no interference is observed for polarizer P_3 oriented at (a) 0° or (c) 90° but high visibility interference fringes are recovered for P_3 oriented at (b) 45°.

equation (2.32) and multiplying the resulting Jones vector with another matrix for a polarizer oriented at ψ, as follows:

$$\varepsilon_{\text{out}} = \begin{pmatrix} \cos^2 \psi & \sin \psi \cos \psi \\ \sin \psi \cos \psi & \sin^2 \psi \end{pmatrix} \frac{1}{4\sqrt{2}} \begin{pmatrix} e^{i\phi} \\ 1 \end{pmatrix}$$
$$= \frac{1}{4\sqrt{2}} \begin{pmatrix} \cos^2 \psi e^{i\phi} + \cos \psi \sin \psi \\ \cos \psi \sin \psi e^{i\phi} + \sin^2 \psi \end{pmatrix}. \tag{2.34}$$

The intensity, in this case, is calculated to be

$$I = \frac{1}{32}(1 + \sin(2\psi)\cos(\phi)), \tag{2.35}$$

which implies that high visibility interference fringes will become visible when $\psi = 45°$ whereas no interference will be observed for $\psi = 0°$ or $\psi = 90°$. For $\psi = 45°$, fringes are observable even when P_1 and P_2 are mutually perpendicular, even though in our earlier experiments without P_3 mutually perpendicular P_1 and P_2 gave us no fringes. The experimental results shown in figure 2.14 agree with this prediction.

But why is this called 'erasure'? In the setup of figure 2.10, we saw that when the polarizers P_1 and P_2 were cross-polarized (horizontal and vertical, respectively), the two paths were 'tagged' and no interference could be observed. Which-path information was available to us. We then added a polarizer oriented at 45° after the second BS (figure 2.13), 'erasing' the which-path information and reinstating interference.

The counter-intuitive aspect of this experiment is the delayed choice [13]. This warrants some discussion. Light, having a finite speed, hits polarizer P_3 seemingly *after* taking one of the two tagged paths. Therefore, common sense has it that just after exiting the second BS, the which-path information is already available to us and interference should not be observable. However, as the light passes through P_3, the available which-path information can be *erased* post facto, and even though we believe that the light had originally taken one of the two paths, we still see interference. A component that is inserted into the apparatus after the interactions with the apparatus are fully completed is modifying what happened earlier—the present is apparently altering the past! At least this is how the popular account of delayed choice and quantum erasure is narrated [14, 15]. We have, however, shown that, surprise aside, the Jones calculus can easily describe the experimental outcomes.

In the later chapters we will see that this phenomenon also applies to *single* photons, implying single-photon interference and quantum erasure! This is where the results are even more astonishing. This erasure experiment is a reconstruction of Wheeler's classic delayed choice experiment [16, 17], which plays a central role in the debate about the foundations of quantum physics.

In the next chapter, we will review the fundamentals of quantum mechanics and we will then begin our experimental investigation of single photons.

References

[1] Hecht E 1998 *Optics* (Reading, MA: Addison-Wesley)
[2] Peatross J and Ware M 2011 *Physics of Light and Optics* (Provo, UT: Department of Physics, Brigham Young University)
[3] Pedrotti F L, Pedrotti L M and Pedrotti L S 2017 *Introduction to Optics* (Cambridge: Cambridge University Press)
[4] Thorne K S and Blandford R D 2017 *Modern Classical Physics* (Princeton, NJ: Princeton University Press)
[5] Fox M 2006 *Quantum Optics: An Introduction* (Oxford: Oxford University Press)
[6] Collett E 2005 *Field Guide to Polarization* (Bellingham, WA: SPIE)
[7] Mansuripur M and Wright E M 2023 Fundamental properties of beamsplitters in classical and quantum optics *Am. J. Phys.* **91** 298–306
[8] Collett E and Schaefer B 2008 Visualization and calculation of polarized light. I. The polarization ellipse, the poincaré sphere and the hybrid polarization sphere *Appl. Opt.* **47** 4009–16
[9] Goldstein D H 2016 *Polarized Light* (Boca Raton, FL: CRC Press)
[10] Waseem M H, Anwar M S *et al* 2019 Hippopedal intensity plots: drawing comparisons between antenna and optical polarimetry *Appl. Opt.* **58** 8442–8

[11] Beck M 2012 *Quantum Mechanics: Theory and Experiment* (Oxford: Oxford University Press)

[12] Dimitrova T L and Weis A 2010 Single photon quantum erasing: a demonstration experiment *Eur. J. Phys.* **31** 625

[13] Ma X, Kofler J and Zeilinger A 2016 Delayed-choice gedanken experiments and their realizations *Rev. Mod. Phys.* **88** 015005

[14] Hobson A 1996 Teaching quantum theory in the introductory course *The Phys. Teach.* **34** 202–10

[15] Hillmer R and Kwiat P 2007 A do-it-yourself quantum eraser *Sci. Am.* **296** 90–5

[16] Wheeler J A 1978 The 'past' and the 'delayed-choice' double-slit experiment *Mathematical Foundations of Quantum Theory* (Amsterdam: Elsevier) pp 9–48

[17] Wheeler J A 1980 Delayed-choice experiments and the Bohr-Einstein dialog *The American Philosophical Society and the Royal Society, Papers* (Philadelphia, PA: American Philosophical Society) pp 9–40

IOP Publishing

Quantum Mechanics in the Single-Photon Laboratory (Second Edition)

Muhammad Sabieh Anwar, Faizan-e-Ilahi, Syed Bilal Hyder and Muhammad Hamza Waseem

Chapter 3

Quantum nature of light

The purpose of this chapter is to present a brief overview of quantum theory, inasmuch as is sufficient to understand the quantum picture of light. We will use the theoretical tools discussed in this chapter throughout the single-photon experiments. While popular textbooks [1–3] on quantum mechanics make an excellent introduction to the subject, we follow an approach similar to particular texts [4–6] because we feel they are well-suited to study photons. Mark Beck's textbook [5] covers many of the experiments discussed in this book. In fact, Beck's work has pioneered the use of photons to demonstrate quantum mechanical phenomena to avid learners of the subject and has no doubt inspired us too. One admission, though. This chapter is not intended as a comprehensive introduction to quantum physics. It only serves as a prelude to understanding the experiments that are the very purpose of this book.

3.1 Quantum mechanical states

In the language of quantum physics, a state provides a complete description of a physical system. Quite simply, we describe the state of a physical system through a state vector, generally denoted as a ket $|\psi\rangle$. This vector resides in a complex vector space endowed with an inner product. This space is called a Hilbert space. The superposition of two quantum states $|\psi_1\rangle$ and $|\psi_2\rangle$ is also a valid state of the same Hilbert space and is called a superposition state. It is expressed as

$$|\psi\rangle = \alpha|\psi_1\rangle + \beta|\psi_2\rangle, \qquad (3.1)$$

where α and β are complex numbers and $|\alpha|^2 + |\beta|^2 = 1$. This equality stems from the need for conserving probability. Its conjugate is described by the dual vector, denoted as the bra $\langle\psi|$, and is given by

$$\langle\psi| = \alpha^*\langle\psi_1| + \beta^*\langle\psi_2|, \qquad (3.2)$$

doi:10.1088/978-0-7503-6315-0ch3

where α^* and β^* represent complex conjugates of α and β, respectively. The bras $\langle\psi_1|$ and $\langle\psi_2|$ live in the dual of the original Hilbert space.

The inner product is used to determine the overlap between two vectors $|\psi\rangle$ and $|\phi\rangle$, and is denoted as

$$\langle\psi|\phi\rangle. \tag{3.3}$$

The inner product of a state with itself is real and positive. A state is said to be normalized if its inner product with itself is one. Two states are said to be orthogonal if their inner product is zero. Two normalized states which are orthogonal to each other are called orthonormal. If the component states $|\psi_1\rangle$ and $|\psi_2\rangle$ in equation (3.1) are orthonormal, the numbers α and β can be computed using the inner product as

$$\langle\psi_1|\psi\rangle = \alpha, \tag{3.4}$$

$$\langle\psi_2|\psi\rangle = \beta. \tag{3.5}$$

For a normalized $|\psi\rangle$ in equation (3.1), we have $\langle\psi|\psi\rangle = |\alpha|^2 + |\beta|^2 = 1$, where $|\alpha|^2$ and $|\beta|^2$ represent the probabilities that the state $|\psi\rangle$ will be measured to be in the states $|\psi_1\rangle$ and $|\psi_2\rangle$, respectively. The complex numbers α and β are hence termed probability amplitudes. The above treatment can be easily extended to n states inside the superposition,

$$|\psi\rangle = \alpha|\psi_1\rangle + \beta|\psi_2\rangle + \gamma|\psi_3\rangle + \cdots + \delta|\psi n\rangle \tag{3.6}$$

where $|\alpha|^2 + |\beta|^2 + |\gamma|^2 + \cdots + |\delta|^2 = 1$. All states that can be represented as state vectors are called pure states, whereas states that cannot be represented as state vectors are called mixed states. They are represented by density matrices and are covered in section 3.6.

3.2 Qubits

A qubit forms the simplest quantum system, having just two basis states. Just as a bit is the fundamental unit of information in computing, a qubit (short for **qu**antum **bit**) is the fundamental unit in quantum computing. However, unlike a classical bit, a single qubit can exist in a superposition of 0 and 1 states[1], usually represented as $|0\rangle$ and $|1\rangle$. Hence, we can write the quantum state corresponding to equation (3.1) as the qubit

$$|\psi\rangle = \alpha|0\rangle + \beta|1\rangle. \tag{3.7}$$

A qubit is a mathematical description of a quantum two-state system. It can be physically realized through any system having two orthogonal states, such as spin-1/2 particles [7, 8] and 2-level atoms [9, 10]. One popular approach is to employ the polarization state of a photon [11–21].

---

[1] We have already introduced superposition states in chapter 1.

In chapter 2, the polarization of light was described by a two-component vector (see table 2.1)

$$\begin{pmatrix} A \\ B\, e^{i\phi} \end{pmatrix} \tag{3.8}$$

also called the Jones vector. We can carry over the same description to a Hilbert space, and identify the first and second terms in the vector as the probability amplitudes attached to the horizontal and vertical polarization states. In figure 2.4, these states are points on the diametrically opposite ends of the S_3 axis. Let us denote these orthogonal states as $|H\rangle$ and $|V\rangle$, also identified sometimes as $|0\rangle$ and $|1\rangle$. Thus, the state in equation (3.7) can then be represented also as a column vector,

$$|\psi\rangle = A|H\rangle + B\, e^{i\phi}|V\rangle = A\begin{pmatrix} 1 \\ 0 \end{pmatrix} + B\, e^{i\phi}\begin{pmatrix} 0 \\ 1 \end{pmatrix} = \begin{pmatrix} A \\ B\, e^{i\phi} \end{pmatrix}. \tag{3.9}$$

where A, B, and φ are real numbers. The dual vector (i.e., bra) will then be represented by the row vector

$$\langle\psi| = \begin{pmatrix} A & B\, e^{-i\phi} \end{pmatrix}. \tag{3.10}$$

Equation (3.9) seems similar to the Jones vector representation of the polarization of classical light, covered in chapter 2. Indeed, the polarization-encoded quantum state representation for *single* photons is analogous to the Jones vector representation of polarization for classical light sources. Hence, in table 3.1 we write the qubit

Table 3.1. Polarization-encoded quantum states and the corresponding state vectors. The angles α, 45°, and −45° are with respect to the horizontal axis of the lab frame of reference. In the last row, A, B, and φ are real numbers.

Polarization	Quantum state	Column vector
Linear at angle α	$|\alpha\rangle$	$\begin{pmatrix} \cos\alpha \\ \sin\alpha \end{pmatrix}$
Horizontal (linear along x axis)	$|H\rangle$	$\begin{pmatrix} 1 \\ 0 \end{pmatrix}$
Vertical (linear along y axis)	$|V\rangle$	$\begin{pmatrix} 0 \\ 1 \end{pmatrix}$
Diagonal (linear at 45°)	$|D\rangle$	$\frac{1}{\sqrt{2}}\begin{pmatrix} 1 \\ 1 \end{pmatrix}$
Antidiagonal (linear at −45°)	$|A\rangle$	$\frac{1}{\sqrt{2}}\begin{pmatrix} 1 \\ -1 \end{pmatrix}$
Left circular	$|L\rangle$	$\frac{1}{\sqrt{2}}\begin{pmatrix} 1 \\ i \end{pmatrix}$
Right circular	$|R\rangle$	$\frac{1}{\sqrt{2}}\begin{pmatrix} 1 \\ -i \end{pmatrix}$
Arbitrary	$|\psi\rangle$	$\begin{pmatrix} A \\ B\, e^{i\phi} \end{pmatrix}$

states identical to the Jones vectors listed in table 2.1. However, there is one crucial subtlety. In the case of classical light, we could say that A is the amplitude ('amount') of horizontally polarized light, while B is the amplitude ('amount') of vertically polarized light. However, while dealing with single photons, we must remember that A and B are not the fractions of horizontal and vertical components of light. Rather, their squared moduli $|A|^2$ and $|B|^2$ determine the probability of finding the horizontal and vertical states, respectively, as a result of an appropriate measurement. From equations (3.4) and (3.5), these probabilities are, respectively, $|\langle H|\psi\rangle|^2 = |A|^2$ and $|\langle V|\psi\rangle|^2 = |B|^2$. The most general way of writing the qubit is in fact

$$|\psi\rangle = \begin{pmatrix} a \\ b \end{pmatrix} \tag{3.11}$$

where a and b are both complex numbers with $|a|^2 + |b|^2 = 1$. In this case, the dual vector becomes

$$\langle \psi| = \begin{pmatrix} a^* & b^* \end{pmatrix}. \tag{3.12}$$

At this juncture, it is worthwhile to step back for a moment and appreciate what we mean by single photons. A light beam comprises numerous smaller entities called photons. The simplest way to obtain these photons is to attenuate a perfectly coherent light beam from a laser so that we get literally one photon at a time. However, this process is cumbersome and does not produce *the* kind of light that is truly quantum. For example, the statistics of detection for the highly attenuated beam of light remains Poissonian, whereas true single photons must show sub-Poissonian statistics (section 3.7). Furthermore, the second-order correlation function (section 4.3) for the attenuated beam of light equals 1, whereas we expect a second-order correlation function close to zero. Therefore, instead of attenuating light, we employ another technique called 'heralded' single-photon generation.

This stream of heralded *single* photons behaves markedly different from a clean, classical beam of light. The heralded single photons show distinct sub-Poissonian statistics as we register their arrival on a detector. Furthermore, these photons cannot be halved or broken down into yet smaller 'grains'. This is markedly different from what can be achieved through classical light impinging on a half-half beam splitter that can literally halve the incident light beam into two sub-beams. This graininess of single photons can be demonstrated in a correlation experiment (described in section 4.3), wherein the second-order correlation function for true single photons will be close to zero. This book deals with the confounding properties of these single photons admitting a quantum description with the uniquely quantum characteristics of superposition, nonlocality, and entanglement.

3.3 Transforming quantum states

In the previous chapter on classical optics, we looked at some commonly employed optical elements. Their functional description and matrix representation carry over for quantum systems. However, the mathematical notation differs slightly in that the matrices are called operators and are identified by placing 'hats' on capital letters.

Polarization-encoded quantum states are transformed through two common types of operations. They are achieved by single-input, single-output polarization manipulation elements, such as a linear polarizer and a wave plate. For example, a linear polarizer carries out a projection operation, which, when coupled with a photodetector, is usually termed a 'measurement'. Consider a beam of polarized single photons hitting a linear polarizer whose axis is at an angle θ with respect to the horizontal. If the incoming photons have a quantum state

$$|\psi_1\rangle = a|H\rangle + b|V\rangle, \tag{3.13}$$

where a and b are complex numbers, a photon may emerge with probability

$$|(\cos\theta\langle H| + \sin\theta\langle V|)(a|H\rangle + b|V\rangle)|^2 = a^2\cos^2\theta + b^2\sin^2\theta. \tag{3.14}$$

If it emerges at all, its quantum state will be

$$|\psi_2\rangle = \cos\theta|H\rangle + \sin\theta|V\rangle. \tag{3.15}$$

A wave plate performs a unitary operation, a transformation that preserves the norm of a state. A unitary operator can be denoted as $\hat{U}$ and is formally defined as

$$\hat{U}\hat{U}^\dagger = \hat{U}^\dagger\hat{U} = \hat{\mathbf{1}}, \tag{3.16}$$

where $\hat{U}^\dagger$ represents the adjoint of the operator and $\hat{\mathrm{I}}$ represents the identity operator which is just the 2 × 2 identity matrix. The unitary action of an operator on a state can be represented as

$$\hat{U}|\psi_1\rangle = |\psi_2\rangle \tag{3.17}$$

whereby the following holds true

$$\langle\psi_2|\psi_2\rangle = \langle\psi_1|\hat{U}^\dagger\hat{U}|\psi_1\rangle = \langle\psi_1|\hat{\mathrm{I}}|\psi_1\rangle = \langle\psi_1||\psi_1\rangle, \tag{3.18}$$

i.e., the norm is preserved by an operator satisfying equation (3.16).

For a series of operations represented as $\hat{O}_1, \hat{O}_2, \dots, \hat{O}_n$, the overall operation $\hat{O}_{\mathrm{eff}}$ can be calculated as

$$\hat{O}_{\mathrm{eff}} = \hat{O}_n \dots \hat{O}_2\hat{O}_1. \tag{3.19}$$

Notice the similarity with equation (2.17).

For qubits expressed in the canonical ($\{|H\rangle, |V\rangle\}$) basis, the operations corresponding to wave plates and polarizers are also represented by 2 × 2 matrices identical to the corresponding Jones matrices. As a convenient reference for the ensuing quantum experiments, we enlist the three most commonly used operators in table 3.2. Note that the operator $\hat{O}_P(\theta)$ representing a linear polarizer is not unitary!

We have briefly hinted at measurement. Let's now explore it in further detail.

Table 3.2. Quantum operators for commonly used polarization manipulating elements. The axis orientations, represented by θ, are with respect to the horizontal axis of the lab frame.

Optical element	Quantum operator	Operator representation in the $\{|H\rangle, |V\rangle\}$ basis
Linear polarizer (axis at θ)	$\hat{O}_P(\theta)$	$\begin{pmatrix} \cos^2\theta & \sin\theta\cos\theta \\ \sin\theta\cos\theta & \sin^2\theta \end{pmatrix}$
Half-wave plate (fast axis at θ)	$\hat{O}_{HWP}(\theta)$	$\begin{pmatrix} \cos 2\theta & \sin 2\theta \\ \sin 2\theta & -\cos 2\theta \end{pmatrix}$
Quarter-wave plate (fast axis at θ)	$\hat{O}_{QWP}(\theta)$	$\begin{pmatrix} \cos^2\theta + i\sin^2\theta & (1-i)\sin\theta\cos\theta \\ (1-i)\sin\theta\cos\theta & \sin^2\theta + i\cos^2\theta \end{pmatrix}$

3.4 Measuring quantum states

Quantum measurement is a rich subject and comes in a variety of shades [22]. Here we discuss its simplest manifestation, of projective measurement, which maximally disturbs the state. In quantum physics, any quantity that can be physically measured is represented by an observable (say O), and to which we can write a corresponding Hermitian operator $\hat{O}$. A Hermitian operator is a matrix with real eigenvalues, and its eigenvectors form an orthonormal basis. This means that we can use the eigenvectors of a Hermitian operator to represent any vector in the Hilbert space. For example, suppose a Hermitiam operator $\hat{O}$ has orthogonal eigenvectors $|\lambda_1\rangle, |\lambda_2\rangle$, up to $|\lambda_n\rangle$. This operator acts on states belonging to the same Hilbert space. Any such state can always be written as

$$|\psi\rangle = \alpha|\psi_1\rangle + \beta|\psi_2\rangle + \gamma|\psi_3\rangle + \cdots + \delta|\psi_n\rangle \tag{3.20}$$

$\{\alpha, \beta, \gamma, \delta\}$ being complex numbers. While measuring O in a system having the state given by $|\psi\rangle$, the probability of obtaining the eigenvalue λ as a result of the measurement is given by $P(\lambda||\psi\rangle) = |\langle\lambda|\psi\rangle|^2$, where $|\lambda\rangle$ is the corresponding eigenstate. This is called Born's rule [5]. If the outcome is λ, then the measurement has created the state $|\lambda\rangle$. This outcome is the essence of a typical strong measurement.

Before making the measurement, the quantum system existed in the state $|\psi\rangle$. However, once the measurement is made, the system is left projecting in the state $|\lambda\rangle$. The act of measurement is, therefore, to change the state of the system. This kind of measurement is called a von Neumann projection[2]. If a subsequent measurement is made in the same basis, the system is always found to be in the same state with 100% probability.

In most of our experiments, the observable is the polarization state. If we are measuring in the canonical basis, our eigenvalues are h and v (we can assign any

---

[2] There could also be many other kinds of measurements, such as weak measurements. However, they do not concern us in this book.

numbers or values to them depending on our apparatus or convention). Generally, we can assign numerical values $h = +1$ and $v = -1$. So, if the state is projected onto $|H\rangle$, the measurement outcome is $h = +1$. If the state is projected onto $|V\rangle$, the measurement outcome is $v = -1$.

If we take a beam of photons generated in the state described by equation (3.11) and place a horizontally oriented linear polarizer in front of the beam, then some fraction of the incoming photons will transmit through. However, a priori, we cannot tell which photon is transmitted and which is blocked. The transmission is random. We can, at best, talk about the probability of the transmission of photons, which is calculated as

$$P(H\,||\psi\rangle) = |\langle H|\psi\rangle|^2 = |a|^2. \tag{3.21}$$

Similarly, if the polarizer is rotated such that its transmission axis is vertically oriented, the probability of transmission will be

$$P(V\,||\psi\rangle) = |\langle V|\psi\rangle|^2 = |b|^2. \tag{3.22}$$

Now, if another horizontally oriented polarizer is placed after the first polarizer, we get no light transmitted because after the first projection the system is left in the state $|V\rangle$ and

$$P(H\,||V\rangle) = |\langle H|V\rangle|^2 = 0. \tag{3.23}$$

On the contrary, if the second polarizer is vertically polarized, we obtain

$$P(V\,||V\rangle) = |\langle V|V\rangle|^2 = 1, \tag{3.24}$$

meaning that all light coming out of the first polarizer will also be transmitted through the second polarizer. We get similar results if we perform similar measurements with a classical light source. However, instead of probabilities, we obtain proportional intensities.

So far, in the quantum description of light, we have replaced Jones vectors with state vectors, Jones matrices with operators, and intensities with probabilities. While polarization optics for a single beam of light does not look very different from quantum and classical world perspectives, two or more particles can show behavior with no classical counterpart. That is the peculiar quantum world! For instance, two particles can become entangled, which is a purely quantum mechanical phenomenon. We now discuss composite systems.

3.5 Composite systems and entangled states

In this section, we extend our discussion from single photons to pairs of single photons and describe this bipartite system in terms of photon polarization. Generally, if we have two independent quantum systems, then we can express their composite state as the direct product given by

$$|\psi\rangle = |\psi_1\rangle \otimes |\psi_2\rangle \equiv |\psi_1\rangle|\psi_2\rangle \equiv |\psi_1\psi_2\rangle. \tag{3.25}$$

These are merely three different ways of expressing the same quantum state. The direct product combines two vectors in different Hilbert spaces to form a larger vector, which expresses the two-particle state in a higher-dimensional Hilbert space. In the case of polarization-encoded photons, the composite system is usually two beams of photons, and the state may be $|HH\rangle, |VV\rangle, |HV\rangle, |VH\rangle$, or superpositions thereof. Some examples of superposition direct product states are

$$|V\rangle \otimes \left(\frac{1}{\sqrt{2}}|H\rangle + \frac{1}{\sqrt{2}}|V\rangle\right) = \frac{1}{\sqrt{2}}|VH\rangle + \frac{1}{\sqrt{2}}|VV\rangle, \tag{3.26}$$

$$|H\rangle \otimes \left(\frac{1}{\sqrt{3}}|H\rangle + i\sqrt{\frac{2}{3}}|V\rangle\right) = \frac{1}{\sqrt{3}}|HH\rangle + i\sqrt{\frac{2}{3}}|HV\rangle, \tag{3.27}$$

$$\left(\frac{1}{\sqrt{2}}|H\rangle + \frac{1}{\sqrt{2}}|V\rangle\right) \otimes \left(\frac{1}{\sqrt{2}}|H\rangle + \frac{1}{\sqrt{2}}|V\rangle\right) = \frac{1}{2}|HH\rangle + \frac{1}{2}|HV\rangle + \frac{1}{2}|VH\rangle + \frac{1}{2}|VV\rangle. \tag{3.28}$$

The enumerated states do not describe all the possible two-photon states. According to the superposition principle, a general composite state can be expressed as

$$|\psi\rangle = \sum_{n_1,n_2} a_{n_1,n_2}|\psi_{n_1}\psi_{n_2}\rangle. \tag{3.29}$$

where complex numbers a_n represent probability amplitudes and $\sum_{n_1,n_2}|a_{n_1,n_2}|^2 = 1$.

Starting with the general form in equation (3.29), we can also work our way backward and attempt to factorize the state into a direct product of component states. For example, consider given the RHS of equation (3.27), one is asked to factorize to yield the representation on the LHS. The interesting point is that this may not *always* be possible. States that are not expressible as direct products of component states are called entangled states [5, 23]. Mathematically, these states may appear to be just run-of-the-mill but, physically, these states manifest some of the most bizarre characteristics of quantum mechanics. We will experimentally play with such systems in the forthcoming chapters. For example, we will generate one of the four famous Bell states [23], which are given by

$$|\Phi^+\rangle = \frac{1}{\sqrt{2}}(|HH\rangle + |VV\rangle), \tag{3.30}$$

$$|\Phi^-\rangle = \frac{1}{\sqrt{2}}(|HH\rangle - |VV\rangle), \tag{3.31}$$

$$|\Psi^+\rangle = \frac{1}{\sqrt{2}}(|HV\rangle + |VH\rangle), \text{ and} \tag{3.32}$$

$$|\Psi^-\rangle = \frac{1}{\sqrt{2}}(|HV\rangle - |VH\rangle). \tag{3.33}$$

These are all entangled states because neither of these states can be expressed as a direct product of two independent qubits. It is not possible to factorize them and assign independent identities to individual particles, which are super-correlated with each other in a strictly quantum sense.

Furthermore, in each Bell state, the two-photon system can be viewed as simultaneously *existing* in two states. For example, $|\Phi^+\rangle$ is in $|HH\rangle$ and $|VV\rangle$ simultaneously. If we measure one of the two photons of $|\Phi^+\rangle$ and find it in the state $|H\rangle$, then the other photon is surely to be found in the state $|H\rangle$, whether or not we choose to measure it. Similarly, if one of the photons is measured in $|V\rangle$, then the other will project or 'collapse' to $|V\rangle$. We randomly project between $|H\rangle$ and $|V\rangle$ for one photon but whatever the case the state measurement by one party instantaneously determines the state of the *other* party. This is the gist of entanglement—random parts making up a nonrandom super-correlated whole!

3.6 Mixed states and the density matrix

So far, all the states that we have described can be represented as state vectors $|\psi\rangle$ and are called pure states. Even the entangled states expressed in equations (3.30)–(3.33) are pure states. No doubt they are superpositions, but they are still quantum mechanically pure. On the contrary, sometimes, the system may be prepared as a random mixture of two possible states (say $|HH\rangle$ and $|VV\rangle$). In this case, the system is in an incoherent *statistical mixture* of the two possible states. For instance, the state $|HH\rangle$ may be added to the mixture with a probability of 50%, and the state $|VV\rangle$ may be introduced into the mixture with a probability of 50%. Such states are called mixed states and cannot be represented as kets. This mixture is unequivocally different from the pure superposition state: $(|HH\rangle + |VV\rangle)/\sqrt{2}$.

Mixed states can be represented using the density matrix formalism [23], in which the density operator $\hat{\rho}$ represents a generalized quantum state. Additionally, this operator can describe both pure *and* mixed states, unlike the state vector which only describes pure states. If a mixture of states $|\psi_j\rangle$ is prepared with respective probabilities $P(|\psi_j\rangle) = p_j$, the density operator can be expressed as

$$\hat{\rho} = \sum_j p_j |\psi_j\rangle\langle\psi_j|. \tag{3.34}$$

The probabilities must be real positive numbers in the range $0 \leqslant p_j \leqslant 1$ and must be normalized $\sum_j p_j = 1$. The states that are part of mixed states need not be orthogonal or form a basis. The component states are any states that the system can be prepared in. For a pure state $|\psi\rangle$, the density operator of equation (3.34) simply reduces to

$$\hat{\rho} = |\psi\rangle\langle\psi|. \tag{3.35}$$

This implies that the density operator of a mixed state is a probability-weighted superposition of density operators of pure states. The normalization condition for a density matrix requires $Tr(\hat{\rho}) = 1$. Just like any other operator for a system with discrete basis, the density operator can be expressed as a matrix.

Let us look at a few examples of density matrices. The density matrix for the pure Bell state given in equation (3.30) can be computed as

$$\hat{\rho}_{\phi^+} = |\phi^+\rangle\langle\phi^+|$$

$$= \frac{1}{\sqrt{2}}(|HH\rangle + |VV\rangle)\frac{1}{\sqrt{2}}(\langle HH| + \langle VV|)$$

$$= \frac{1}{2}|HH\rangle\langle HH| + \frac{1}{2}|HH\rangle\langle VV| + \frac{1}{2}|VV\rangle\langle HH| + \frac{1}{2}|VV\rangle\langle VV| \quad (3.36)$$

$$= \frac{1}{2}\begin{pmatrix} 1 & 0 & 0 & 1 \\ 0 & 0 & 0 & 0 \\ 0 & 0 & 0 & 0 \\ 1 & 0 & 0 & 1 \end{pmatrix}.$$

The matrix form in the last line can be written with the help of basic matrix algebra. For example, using the vector forms of $|H\rangle$ and $|V\rangle$ in table 3.1, we can write the vector $|HH\rangle$ as

$$|HH\rangle = \begin{pmatrix} 1 \\ 0 \end{pmatrix} \otimes \begin{pmatrix} 1 \\ 0 \end{pmatrix} = \begin{pmatrix} 1 \\ 0 \\ 0 \\ 0 \end{pmatrix} \quad (3.37)$$

with the dual bra

$$\langle HH| = \begin{pmatrix} 1 & 0 & 0 & 0 \end{pmatrix}. \quad (3.38)$$

Finally, juxtaposing the bra and the ket and multiplying the row and column vectors yields

$$|HH\rangle\langle HH| = \begin{pmatrix} 1 & 0 & 0 & 0 \\ 0 & 0 & 0 & 0 \\ 0 & 0 & 0 & 0 \\ 0 & 0 & 0 & 0 \end{pmatrix}. \quad (3.39)$$

This process can be repeated for $|HH\rangle\langle VV|$, $|VV\rangle\langle HH|$, and $|VV\rangle|VV\rangle$ to finally construct the matrix in equation (3.36).

On the contrary, the density matrix for a mixed two-photon state, in which half the time the photons are detected in the state $|HH\rangle$ and half the time they are detected in the state $|VV\rangle$, is given by

$$\hat{\rho}_{\text{mix}} = \frac{1}{2}|HH\rangle\langle HH| + \frac{1}{2}|VV\rangle\langle VV|$$

$$= \frac{1}{2}\begin{pmatrix} 1 & 0 & 0 & 0 \\ 0 & 0 & 0 & 0 \\ 0 & 0 & 0 & 0 \\ 0 & 0 & 0 & 1 \end{pmatrix}. \quad (3.40)$$

This density matrix $\hat{\rho}_{\text{mix}}$ is clearly distinct from $\hat{\rho}_{\phi^+}$, due to the absence of the cross-diagonal entries on the extreme corners. These cross-diagonal terms are a hallmark of a coherent superposition. In the mixed state $\hat{\rho}_{\text{mix}}$, the admixture of states is incoherent and is like jumbling states together. There could be an infinite variety of mixed states. Another important two-qubit state is the Werner state [24], represented by

$$\hat{\rho}_W = \varepsilon|\psi_{\text{ent}}\rangle\langle\psi_{\text{ent}}| + (1-\varepsilon)\frac{\hat{1}}{4}, \tag{3.41}$$

where $|\psi_{\text{ent}}\rangle\langle\psi_{\text{ent}}|$ represents the density matrix of a maximally entangled state, given by one of the four Bell states described in the previous section, and $\hat{1}/4$ (where $\hat{1}$ is the 4×4 identity matrix) represents the density matrix of maximally mixed two-qubit state. The variable ε represents the probability that the photons are in the state $|\psi_{\text{ent}}\rangle\langle\psi_{\text{ent}}|$ and $(1-\varepsilon)$ represents the probability the photons are in the maximally mixed state $\hat{1}/4$.

For a pure state given by $|\psi\rangle\langle\psi|$, it is evident that $Tr(\hat{\rho}^2) = Tr(\hat{\rho}) = 1$. On the contrary, for mixed states, it can be seen that $\hat{\rho}^2 \neq \hat{\rho}$ and $Tr(\hat{\rho}^2) < 1$. So, $Tr(\hat{\rho}^2)$ can be used as a figure of merit for the 'purity' of a system. The closer it is to 1, the purer the state is [5]. For a completely mixed two-qubit state, $Tr(\hat{\rho}^2) = 1/4$. The denominator is $4 = 2^2$, which is the dimensionality of the system.

As a last example, consider an equal mixture of the states $|HH\rangle$, $|HV\rangle$, $|VH\rangle$, and $|VV\rangle$. The density matrix of this incoherent mixture will be

$$\frac{1}{4}|HH\rangle\langle HH| + \frac{1}{4}|HV\rangle\langle HV| + \frac{1}{4}|VH\rangle\langle VH| + \frac{1}{4}|VV\rangle\langle VV| = \frac{\hat{1}}{4} \tag{3.42}$$

which is maximally mixed. Each of the constituents of the mixture is a separable state of the two photons. Interestingly, we can also show that we could still obtain the maximally mixed state if we compose an equal mixture of non-separable, entangled Bell states. Try to see if you can verify the following:

$$\frac{1}{4}|\Phi^+\rangle\langle\Phi^+| + \frac{1}{4}|\Phi^-\rangle\langle\Phi^-| + \frac{1}{4}|\Psi^+\rangle\langle\Psi^+| + \frac{1}{4}|\Psi^-\rangle\langle\Psi^-| = \frac{\hat{1}}{4}. \tag{3.43}$$

The last example shows that incoherently mixing entangled states can lead to a separable state.

3.7 Photon statistics

Apart from multi-photon entangled systems, another aspect in which single photons remarkably differ from coherent light is in the statistics of photodetection. The photodetection statistics for coherent light with a stable intensity, such as a laser, are described by a Poisson distribution [4], represented as

$$p(N) = \frac{\bar{N}^N}{N!}e^{-\bar{N}} \quad \text{with} \quad N = 0, 1, 2, \dots, \tag{3.44}$$

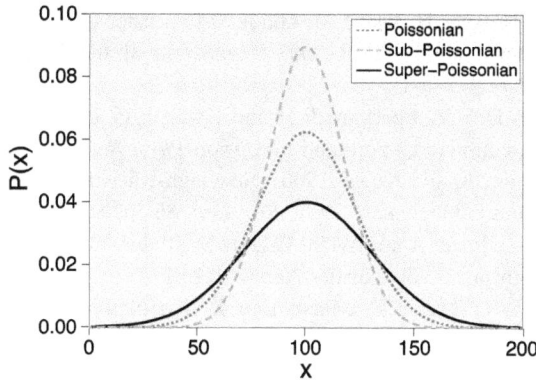

Figure 3.1. Comparison of Poissonian, sub-Poissonian, and super-Poissonian probability distributions for 100 mean photodetections ($\bar{N} = 100$). The sub-Poissonian distribution has the smallest spread, while the super-Poissonian distribution has the largest.

where N represents the number of photons successfully detected in a given time duration, $\bar{N}$ represents the mean photodetections, while the standard deviation is given by $\Delta N = \sqrt{\bar{N}}$. Light from a thermal source, such as a filament lamp, presents a larger standard deviation due to thermal fluctuations, resulting in super-Poissonian statistics ($\Delta N > \sqrt{\bar{N}}$). Photons produced by a thermal source arrive in clusters at the detectors, and this kind of light is labeled *bunched*.

In contrast, a stream of single photons displays sub-Poissonian statistics ($\Delta N < \sqrt{\bar{N}}$) as a result of temporally distinct photodetections. This light is called *antibunched* and lacks a classical counterpart. The frequency of detections for these three kinds of light is shown as bell-like curves in figure 3.1.

The theoretical toolbox developed in this chapter will prove useful throughout one's journey through this book. In the next chapter, we will discuss how to produce single photons through the process of spontaneous parametric downconversion and describe a signature experiment that verifies the particle nature of photons.

References

[1] Griffiths D J 2016 *Introduction to Quantum Mechanics* (Cambridge: Cambridge University Press)

[2] Shankar R 2012 *Principles of Quantum Mechanics* (Berlin: Springer)

[3] Dirac P A M 1981 *The Principles of Quantum Mechanics* (Oxford: Oxford University Press)

[4] Fox M 2006 *Quantum Optics: An Introduction* (Oxford: Oxford University Press)

[5] Beck M 2012 *Quantum Mechanics: Theory and Experiment* (Oxford: Oxford University Press)

[6] Lvovsky A I 2018 *Quantum Physics: An Introduction Based on Photons* (Berlin: Springer)

[7] Cory D G, Fahmy A F and Havel T F 1997 Ensemble quantum computing by NMR spectroscopy *Proc. Natl. Acad. Sci. U.S.A.* **94** 1634–9

[8] Weinstein Y S, Pravia M A, Fortunato E M, Lloyd S and Cory D G 2001 Implementation of the quantum Fourier transform *Phys. Rev. Lett.* **86** 1889

[9] Monroe C 2002 Quantum information processing with atoms and photons *Nature* **416** 238

[10] Schmidt-Kaler F, Häffner H, Riebe M, Gulde S, Lancaster G P T, Deuschle T, Becher C, Roos C F, Eschner J and Blatt R 2003 Realization of the Cirac-Zoller controlled-not quantum gate *Nature* **422** 408

[11] White A G, James D F V, Eberhard P H and Kwiat P G 1999 Nonmaximally entangled states: production, characterization, and utilization *Phys. Rev. Lett.* **83** 3103

[12] Sanaka K, Kawahara K and Kuga T 2001 New high-efficiency source of photon pairs for engineering quantum entanglement *Phys. Rev. Lett.* **86** 5620

[13] Mair A, Vaziri A, Zeilinger A and Weihs G 2001 Multi dimensional photon entanglement of quantum states with phase singularities *Nature* **412** 3123

[14] Nambu Y, Usami K, Tsuda Y, Matsumoto K and Nakamura K 2002 Generation of polarization-entangled photon pairs in a cascade of two Type-I crystals pumped by femtosecond pulses *Phys. Rev.* A **66** 033816

[15] Yamamoto T, Koashi M, Özdemir Ş K and Imoto N 2003 Experimental extraction of an entangled photon pair from two identically decohered pairs *Nature* **421** 343

[16] Sergienko A V, Di Giuseppe G, Atatüre M *et al* 2003 Entangled-photon state engineering *Proc. 6th Int. Conf. on Quantum Communication, Measurement and Computing (QCMC)* (Rinton Princeton) pp 147–52

[17] Pittman T B, Fitch M J, Jacobs B C and Franson J D 2003 Experimental controlled-not logic gate for single photons in the coincidence basis *Phys. Rev.* A **68** 032316

[18] O'Brien J L, Pryde G J, White A G, Ralph T C and Branning D 2003 Demonstration of an all-optical quantum controlled-not gate *Nature* **426** 264

[19] Marcikic I, De Riedmatten H, Tittel W, Zbinden H and Gisin N 2003 Long-distance teleportation of qubits at telecommunication wavelengths *Nature* **421** 509

[20] Pearson B J and Jackson D P 2010 A hands-on introduction to single photons and quantum mechanics for undergraduates *Am. J. Phys.* **78** 471–84

[21] Brody J and Selton C 2018 Quantum entanglement with Freedman's inequality *Am. J. Phys.* **86** 412–6

[22] Jordan A N and Siddiqi I A 2024 *Quantum Measurement: Theory and Practice* (Cambridge: Cambridge University Press)

[23] Nielsen M A and Chuang I 2010 *Quantum Computation and Quantum Information* (Cambridge: Cambridge University Press)

[24] Werner R F 1989 Quantum states with Einstein-Podolsky-Rosen correlations admitting a hidden-variable model *Phys. Rev.* A **40** 4277

IOP Publishing

Quantum Mechanics in the Single-Photon Laboratory
(Second Edition)

Muhammad Sabieh Anwar, Faizan-e-Ilahi, Syed Bilal Hyder and Muhammad Hamza Waseem

Chapter 4

Experiments related to generating single photons

Experiments based on statistics of photons and single-photon states have been explored as an effective tool to study and teach quantum mechanics in the instructional laboratory, where simplicity, affordability, and modularity are important concerns. Important work in this regard has been championed by experimental physics groups headed by Beck [1–4], Galvez [1, 2, 5–7], and Lukishova [8–10], and also surveyed in chapter 1.

The content of this manuscript is inspired by all of these studies, especially Beck's work [1–4]. A general overview of this 'single-photon laboratory' can be visualized in figure 4.1, which schematically depicts three major components of the laboratory: the optical setup, photon counting, and post-processing of photodetection statistics. Furthermore, figure 4.2 shows a bird's-eye view of the laboratory constructed and housed in PhysLab, LUMS, Pakistan.

We will refer to our tabletop single-photon experiments through the following nomenclature:

- Q1: Spontaneous parametric downconversion
- Q2: Testing the particle-like behavior of light
- Q3: Estimating the polarization state of single photons
- Q4: Visualizing the polarization state of single photons
- NL1: Freedman's test of locality
- NL2: The Clauser–Horne–Shimony–Holt (CHSH) test of locality
- NL3: Hardy's test of locality
- Q5: Single-photon interference and quantum eraser
- Q+NL: Nonlocal quantum erasure
- QST: Quantum state tomography

This chapter deals with Q1 and Q2 only, whereas Q3 onward will be the subject of subsequent chapters. However, at this point, we will highlight the instrumentation commonly employed in all of the experiments.

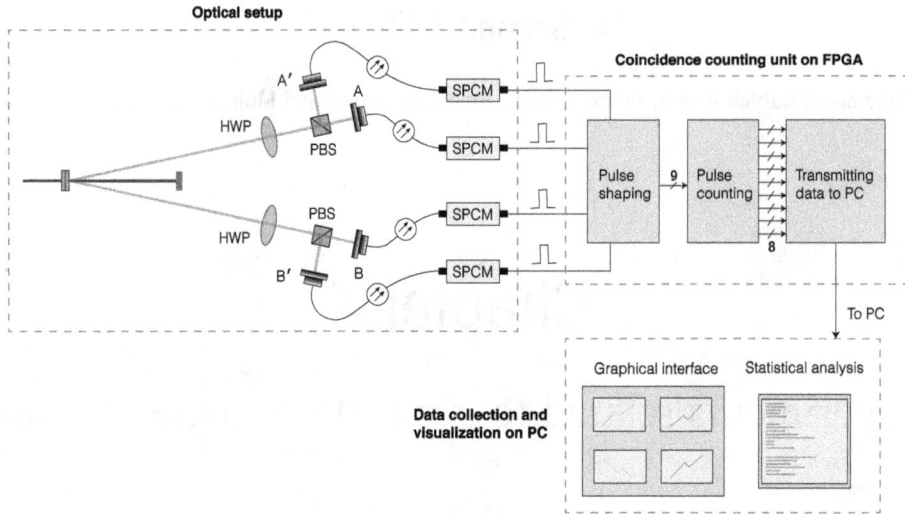

Figure 4.1. The three major compartments of the single-photon experiments: the optical setup, the photon counting mechanism, and the post-processing system of the photon statistics. The elements up till the SPCMs form the optical setup; SPCMs detect the photons and generate electrical signals. The FPGA detects and counts single-photon and coincidence photon pulses, and then sends the photocounts periodically via serial communication to a Python program running on a computer.

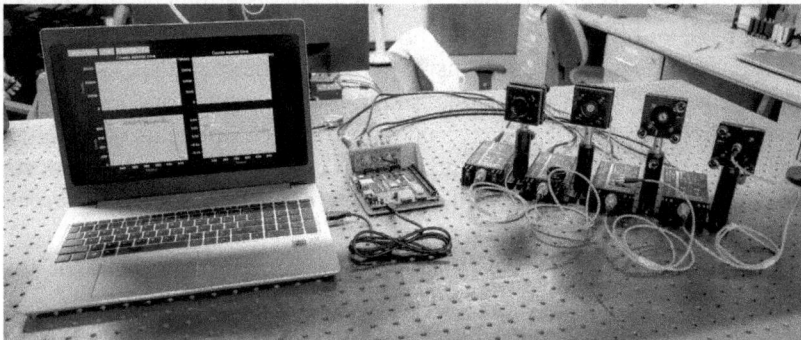

Figure 4.2. The single-photon laboratory in PhysLab, LUMS. The optical components are set up on the optical table. Single-photon detectors transmit TTL pulses for each photodetection. The pulses are counted by an FPGA, which transmits photon count information to a PC, where counts are monitored via a Python program. The single-photon experiments are performed with the lights turned off.

Precisely speaking, experiment Q1 is about setting up a source for creating single photons using the process of downconversion. Experiment Q2 shows the granular nature of light. Often dubbed as evidence for the existence of photons, this experiment serves as a confirmation of the successful generation of single photons. Experiments Q3 and Q4, respectively, estimate and visualize the polarization state of single photons. Experiment Q5 brings home aspects of wave-particle duality by exploring the interference of single photons and quantum erasure, while Q+NL combines interference with nonlocal effects.

The experiments are modular. In this book, we have sequenced them such that subsequent experiments build up on the preceding ones. We begin by describing the instrumentation that forms the backbone of all of the experiments.

4.1 General components of the lab

For convenience, we divide our experimental setup into three groups as shown in figure 4.1: the optical setup, the coincidence counting unit, and the data processing on PC. A detailed listing of the components is also furnished in appendix B.

4.1.1 Optical setup

The optical setup comprises all the equipment that is concerned with the generation, polarization manipulation, or detection of photons. In this section, we mainly discuss the apparatus that produces, collects, and detects single photons.

4.1.1.1 Light source

Pairs of photons are produced through a nonlinear process called downconversion, which will be discussed in sufficient detail in experiment Q1. Downconversion is a highly inefficient process, so the experiments require a sufficiently bright light source. We use a vertically polarized, 405 nm violet-blue laser for all of the experiments described in this book. The laser has a nominal power of 50 mW, which can be controlled by rotating a knob (figure 4.3a). Before falling onto the downconversion crystal, the laser is made to pass through a half-wave plate (HWP), which helps to adjust the laser polarization.

Downconversion is achieved in two stacked β-barium borate (BBO, BaB_2O_4) crystals, cut for type-I downconversion (figure 4.3b). Since they are hygroscopic, the crystals are protected from moisture by placing silica gel beneath them and maintaining a gentle flow of nitrogen over the crystal mount. Moreover, when the crystals are not being used, they are stored in a desiccant jar.

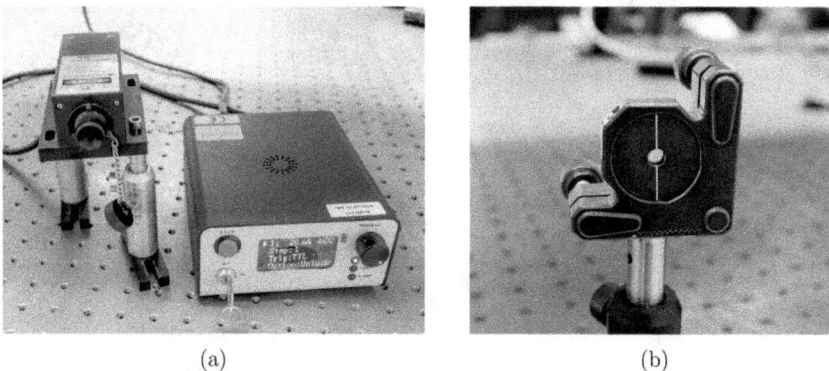

(a)   (b)

Figure 4.3. (a) A 405 nm continuous wave laser (MDL-III-405-50 mW, CNI) is used as the pump laser for the single-photon experiments. (b) Mounted BBO crystals (PABBO5050-405(I)-HA3, Newlight Photonics). The white line shows the axis of one of the crystals.

The performance or quality of the light source can be gauged in terms of single and coincidence count rates, which can be measured in counts per second (cps). Our detection rates can go up to 200 000 for single photons in one beam and up to ≈15 000 for coincidences between the two photon beams.

4.1.1.2 Light collection

The photons are collected and fed into detectors through suitable components. The collection optics help in alignment, as well as rejection of background. The collection optics and detection schemes described in this section will be used identically for all the detectors in all of the experiments.

As shown in figure 4.4, the downconverted photons are collected with fiber-coupling (FC) lenses, coupled into multimode fiber optic cables having FC connectors at both ends and directed into single-photon counting modules where they are counted. Optical fibers help make the system flexible and efficient. In front of the collection lenses are long-pass optical filters. These filters block light having wavelengths smaller than 780 nm. They block ambient or spuriously scattered light, which can not only distort the photon statistics but can also damage the single photon counting module SPCM.

Kinematic mounts hold both the fiber-coupled collimation lenses and the long-pass filters. The mounts are flexible and adjustable, helping in alignment. They are equipped with knobs that can adjust the coupling lenses' horizontal and vertical tilts.

4.1.1.3 Light detection

The fiber takes the photons to the SPCMs, which are a crucial and perhaps the most expensive component in our laboratory. We use SPCMs (figure 4.6) with a single input channel. Inside the SPCM, the input photon is taken to an avalanche photodiode, which can detect photons in our desired wavelength range. The

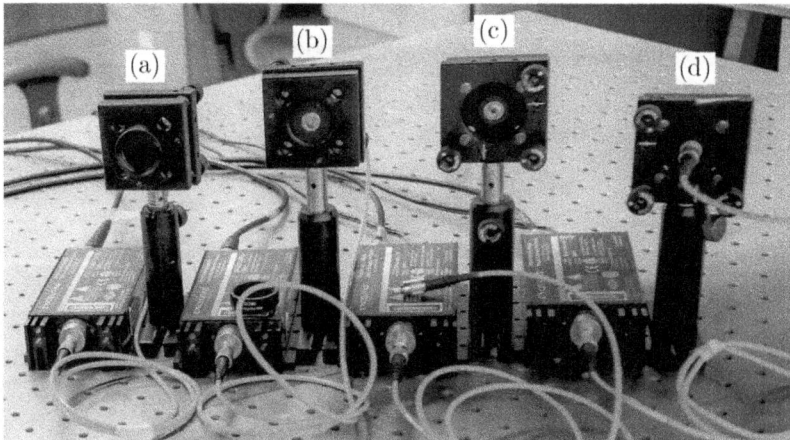

Figure 4.4. The collimators, held in kinematic mounts, collect the downconverted photons. These photons are directed into multimode fiber optic cables that take them to the SPCMs. (a) The mount has a long-pass filter installed. (b) The filter is not installed and the collimator's lens is exposed. (c) The backside of a collimator with the fiber optic disconnected. (d) The backside of a collimator with the fiber optic connected.

Table 4.1. Dark count rates registered for the three kinds of detectors used in the various experiments. All rates are in units of counts per second (cps), and the dead times and pulse widths are in nanoseconds (ns). Assuming Poissonian statistics, the uncertainties are estimated by the square root of the count rates. The dead times and nominal pulse widths are quoted from the data sheets provided by the manufacturer.

Detector	1	2	3	4	Dead time	Pulse width
SPCM-AQ4C	352 ± 19	242 ± 16	263 ± 15	339 ± 18	50	25
SPCM-AQRH	152 ± 12	102 ± 10			22	10
SPCM-EDU			1624 ± 40	815 ± 26	22	10

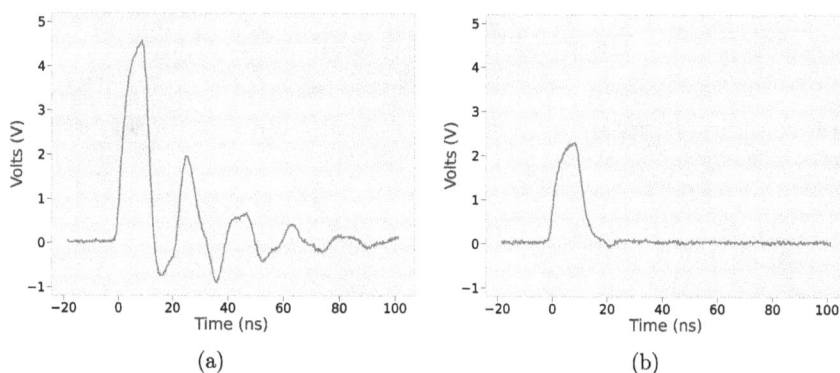

Figure 4.5. (a) Raw output pulse from the SPCM. (b) Signal after impedance matching. These pulses have been captured by a high-speed oscilloscope.

detectors also have a dead time, shown in table 4.1 (dependent on the detector's model). The avalanche photodiode (APD) modules are optimized for a peak photon detection efficiency of about 50% at wavelength 650 nm.

For each photodetection, an output pulse of peak height 5 V and duration 10–25 ns is produced at the Bayonet Neill–Concelman (BNC) output port of the corresponding channel. A high-speed oscilloscope can be used to view this pulse. A typical pulse waveform is shown in figure 4.5a. It is also important to keep the detectors safe from receiving ambient light; otherwise, they can be damaged due to excess photons. Therefore, experiments are conducted with the room's lights turned off. Moreover, thermal fluctuations can cause the APDs to register photons even when there is no light incident on the APDs. These are called dark counts. We used three kinds of detectors and measured their dark counts, which are tabulated in table 4.1.

4.1.2 Coincidence counting unit

Coincidence counting refers to counting simultaneous detections of two or more particles at different detectors [11]. Widely employed in experimental physics, this

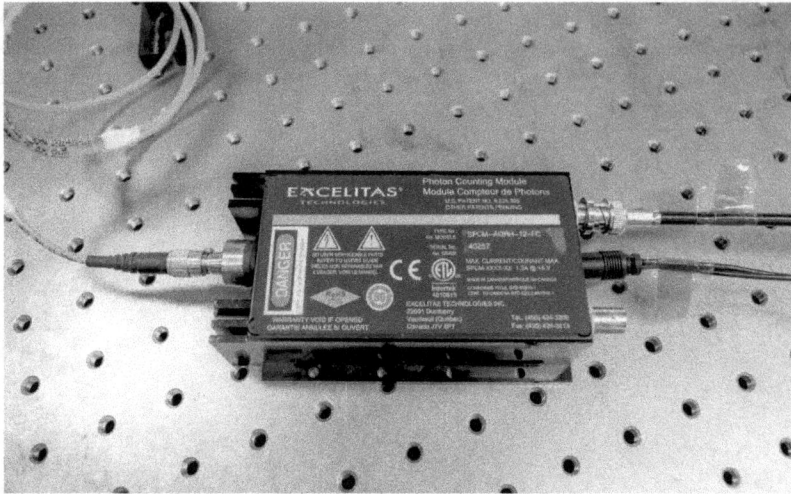

Figure 4.6. Photograph of a single-photon counting module (an SPCM-AQRH, Excelitas). This is one of the four single-photon detectors in our setup. Photons are input from the fiber optic's FC connector, and the detection pulses are transmitted via the BNC terminals at the back.

technique plays an essential role in experiments pertaining to quantum optics. For example, coincidence detection and photon counting lie at the heart of investigating and making use of the quantum characteristics of correlated light sources. As we will see throughout this book, most of these experiments demand counting of only two-fold coincidences, though one experiment (Q2) does require coincidences for more than two detectors [12–14].

Traditionally, coincidence counting has been performed using time-to-amplitude converters (TACs), where each TAC allows to count one coincident photon pair [15]. A TAC has a 'start' and a 'stop' input. If a photon arrives on the 'start' input, it starts a stopwatch, which terminates when another photon hits the 'stop' input. The time interval between the start and stop is registered. If this interval $\tau < \Delta t$, a threshold, we count this as a coincidence. These coincidences are counted and Δt is called the coincidence window. However, coincidence counting can become laborious and costly for multiphoton experiments where correlations between many photons arriving in arbitrary sequences need to be quantified. Moreover, the maximum rate of TAC-based coincidence counting is severely capped by the conversion time involved in each start or stop event. In the last two decades, these problems have been largely addressed and many solutions have been proposed, also focusing on particular applications, such as quantum information processing [16], fluorescence measurements [17, 18], x-ray microscopy [19], and physics education [11, 20–22].

To perform quantum physical experiments, a cost-effective solution for coincidence counting is based on the field-programmable gate array [23, 24]. We discuss our field programmable gate array (FPGA)-based coincidence counting unit (CCU) briefly in this section, whereas details are provided in appendix A.

4.1.2.1 The counting mechanism

The pulses generated by the SPCM are counted by a CCU programmed on an FPGA development board, a Nexys A7 in our case. An FPGA is a chip with integrated circuitry containing many logic blocks that can be programmed to perform the desired operations. The FPGA is embedded in a development board containing various peripherals, such as connectors, switches, buttons, etc. The board is shown in figure 4.7. We use an FPGA with an internal clock of 100 MHz, and our implementation provides us with an effective coincidence window as low as 3 ns. For some experiments, we employ larger coincidence windows of 20 and 40 ns. Hence, it provides an efficient and cost-effective solution to simultaneously count single and coincident photons from several photodetectors.

Using the high-speed clock from the onboard oscillator, the CCU monitors the input channels and counts the pulses and their coincidences. Incoming pulses are considered coincident if they arrive at two or more different channels of the FPGA within a specified coincidence time window Δt (i.e. 3 ns in our case). Coincidence detections are monitored by logically ANDing together the required pulses. The CCU implemented on the FPGA communicates all coincidence and single count information to a computer through serial communication using the UART communication protocol. Finally, the data can be received on the computer by establishing a serial connection port.

Several options are available to open a serial port on the computer, e.g., Matlab, Python, and LabView. For the experiments discussed in this book, we used a program that uses Python's *pySerial* library to establish a serial port connection with the FPGA. This program has an interactive graphical user interface (GUI) that is made using the *matplotlib* library. Using this GUI, the experimenter can not only

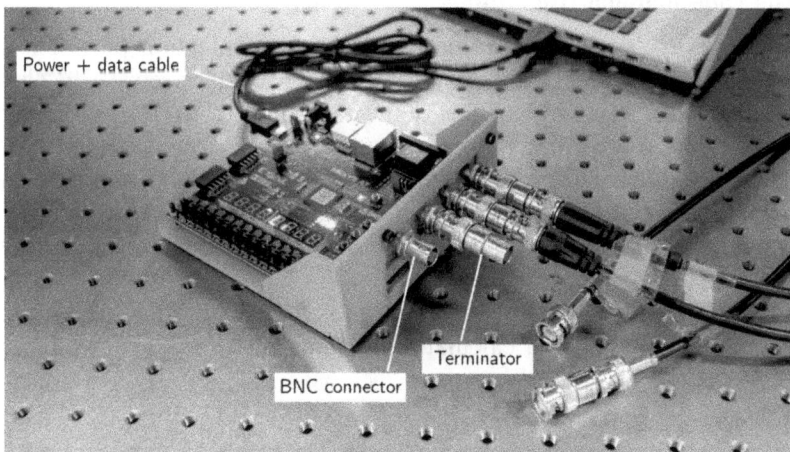

Figure 4.7. The CCU is programmed on an Artix 7 FPGA chip embedded on a Nexys A7 development board. With the current circuitry, it can handle four inputs and measure their coincidences. The BNC connectors have 50 Ω terminators for impedance matching. The micro-USB to USB cable supplies power to the FPGA and also handles data transfer to the PC.

monitor the single detector counts and coincidence counts in live plots but can also easily record the data for specific time intervals.

The input pins of the FPGA are rated 3.3 V, and a 5 V pulse can damage the FPGA permanently, so we should be cautious of what we feed the FPGA through the pins. The raw signal from the SPCMs without impedance matching is shown in figure 4.5a. The signal has a lot of ringing, and the peak voltage reaches as high as 5 V. Not only is this peak voltage unsafe for the FPGA but the ringing can also cause the CCU to count the same pulse more than once. To address these issues, we match the impedance of the BNC's and the receiving ends of the FPGA using 50 Ω terminators, which also function as voltage dividers to rescale the voltage levels. The outcome of impedance matching is a square-like pulse with a peak voltage of about 2.2 V, as shown in figure 4.5b.

4.1.2.2 The communication protocol

For the experiments discussed in this book, we need to use a maximum of four detectors at a time, so our CCU is currently designed to cater to four input channels.

Counts for the detection are each stored in 8-bit counters. In every communication cycle, the FPGA sends 10 8-bit packets (nine counts and one header) to the PC. These packets are transmitted one by one in the order illustrated by figure 4.8. For any experiment, the experimenter may use only the required counts and discard the rest.

Data is communicated to the PC using UART communication protocol via a B-micro-USB to USB Type-A cable. A communication cycle takes 25 μ s (a total of 40 000 cycles per second), and pulse counts are reset after each cycle. We use a Python program developed by the authors of this book to monitor the photocounts.

4.1.3 Data collection and visualization

The PC receives data packets through a serial communication port. The Python program looks for a header and then starts reading packets in sets of 10. The header is discarded from each set, and the rest of the data are used to display the single and coincidence counts in the form of numbers and plots. The front panel of the program (figure 4.9) has a `Stop` button that safely closes the communication port and exits the program, a `Capture Screen` button that saves the current screen in `.jpg` file format, and an `Acquire Data` button that collects the data for the specified time interval and saves it in `.txt` file format.

The files are saved in the same directory as the parent Python file, and the data can be easily accessed later for calculations or data processing. All of the software

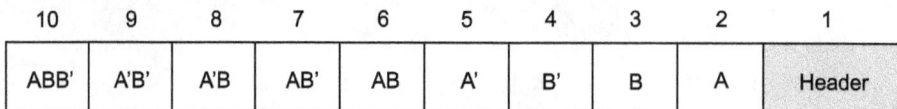

10	9	8	7	6	5	4	3	2	1
ABB'	A'B'	A'B	AB'	AB	A'	B'	B	A	Header

Figure 4.8. In a communication cycle, a header is sent first, followed by all the data register values in a specific order. The header has a constant value and helps keep track of the packet order.

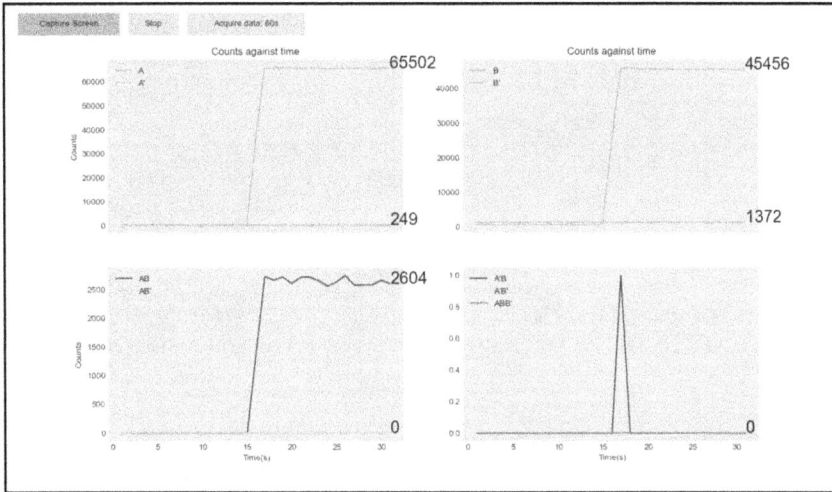

Figure 4.9. A screenshot of the front panel captured using the CAPTURE SCREEN button.

and code can be downloaded from http://www.physlab.org/qmlab. Let us now move on to our first experiment, which generates single photons.

4.2 Q1: Spontaneous parametric downconversion

At the heart of all our single-photon quantum experiments is a process called spontaneous parametric downconversion (SPDC) [20, 25–27]. In SPDC, an electron is excited by the incident pump photon. As the electron returns to the ground state, it emits two photons rather than one. The process is called downconversion because the frequency of the output beams is lower than the frequency of the input beam. This is a nonlinear process because it changes the frequency of the light beam, in contrast to the linear optical processes that we studied in chapter 2, which changes many properties, such as intensity, phase, or polarization but not the frequency.

In SPDC, schematically illustrated in figure 4.10, a pump (p) photon of one frequency produces two photons (signal (s) and idler (i)) of half the original frequency. The downconverted photon pairs are used as the source to study the peculiar quantum nature of light. The process is dubbed 'spontaneous' because the photons are downconverted continuously at random, and we cannot control it with any external stimulus. Furthermore, SPDC is a parametric process because it depends on the electric fields (polarizations) rather than the intensities of the light beams. Therefore, a definite phase relationship exists between the pump and downconverted beams. The phase-matching requirement is outlined in the next section.

SPDC has several merits that make it an effective source of photons. Although the process of SPDC itself is extremely inefficient, it is more efficient than the atomic cascade method used by Grangier *et al* [28] and is, of course, much easier to implement. The downconverted photons are always emitted in pairs. Thus, the detection of one photon tags the presence of its sibling. This allows SPDC to be used as a *heralded* single-photon source. Owing to its relative simplicity, SPDC has been

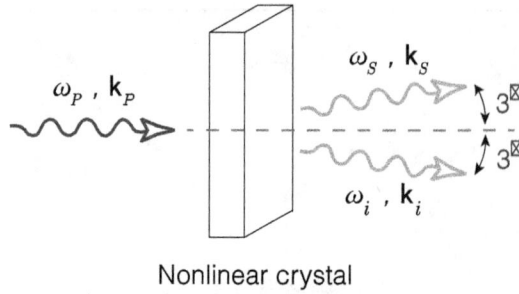

Nonlinear crystal

Figure 4.10. Concept diagram of type-I spontaneous parametric downconversion. A photon of frequency ω_p and wave vector $\mathbf{k}_p$ is absorbed, and two photons of frequencies ω_s and ω_i and wave vectors $\mathbf{k}_s$ and $\mathbf{k}_i$ are released.

used in correlated-photon pair experiments, such as tests of existence of photons [25], single-photon interference [3], quantum erasure [29, 30], state measurement of single photons [3], quantum state tomography [31], and tests of Bell inequalities [32–34].

4.2.1 The downconversion crystal and phase-matching

For SPDC, we used two stacked BBO crystals, each of dimensions $5 \times 5 \times 0.5$ mm. They are stacked up in a way that the axis of one is rotated 90° relative to the other. This allows for downconversion of both horizontally and vertically polarized photons. The BBO crystals are cut for type-I SPDC to produce photon pairs with linear polarization parallel to each other but orthogonal to the polarization of the input beam. This means that the downconversion relations for the two crystals can be described separately as

$$|V\rangle \longrightarrow |HH\rangle, \tag{4.1}$$

$$|H\rangle \longrightarrow |VV\rangle. \tag{4.2}$$

Moreover, if the pump polarization is oriented at an angle θ with respect to the vertical, the vertical and horizontal components of the input beam will, respectively, result in $\cos\theta|HH\rangle$ and $\sin\theta|VV\rangle$. At this stage, we are only looking at the two downconversion crystals separately. However, when we are interested in the polarization correlations of the two beams, an important subtlety arises about this assembly of two BBO crystals [26]. If one cannot distinguish between the photon pairs produced by the two crystals, then the two-photon state is entangled. We will discuss these baffling concepts in more detail in chapter 6, where we talk about entanglement and nonlocality.

The direction the output photons take is determined by the angle formed by the optic axis[1] of the crystal with the direction of propagation of the pump beam (illustrated in figure 4.11). This angle is called the phase-matching angle θ_m. The

---

[1] Optical axis is defined as the direction of propagation through a birefringent material in which the input light experiences a constant refractive index, irrespective of its polarization.

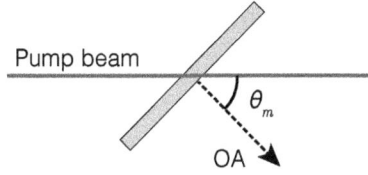

Figure 4.11. Optical axis (OA) of the material is represented by the vector normal to its surface. The angle that the pump beam makes with the optical axis is the phase-matching angle θ_m.

crystal needs to be cut precisely at θ_m for the whole scheme to work properly, and this angle needs to be specified at the time of crystal growth. The manufacturers make this cut very well.

4.2.1.1 Calculations for phase-matching

As illustrated in figure 4.12, the law of conservation of energy requires that

$$\omega_p = \omega_s + \omega_i, \tag{4.3}$$

whereas conservation of momentum requires that

$$\mathbf{k}_p = \mathbf{k}_s + \mathbf{k}_i. \tag{4.4}$$

The frequencies and wave vectors of the optical beams are not independent of each other but are, in fact, related. For the pump wave, for instance, we have the relation

$$k_p = \frac{n_p \omega_p}{c}, \tag{4.5}$$

where n_p denotes the refractive index of the downconversion crystal at the pump frequency. A dispersion relationship correlates refractive index with frequency, $n(\omega)$. Similar dispersion relations exist for signal and idler waves. Therefore, given the angles θ_p, θ_s, and θ_i of the pump, signal, and idler beams with respect to the pump propagation direction, equations (4.4) and (4.5) lead to the conditions

$$n_p \omega_p = n_s \omega_s \cos \theta_s + n_i \omega_i \cos \theta_i \text{ and} \tag{4.6}$$

$$0 = n_s \omega_s \sin \theta_s + n_i \omega_i \sin \theta_i. \tag{4.7}$$

The downconverted photons come out of the crystal at a range of wavelengths and angles. However, for our experiments, we only consider the photons which come out in the horizontal plane and for which $\omega_s = \omega_i = \omega_p/2$, $n_s = n_i$ and $\theta_s = \theta_i$. Therefore, equation (4.6) becomes

$$n_p = n_s \cos \theta_s. \tag{4.8}$$

It is not possible to satisfy this equation in an isotropic medium[2] because, for normal dispersion, $n_p > n_s$ when $\lambda_p < \lambda_s$, leading to unphysical values of the cosine function. However, this problem can be overcome with a BBO crystal, a uniaxial

---

[2] A medium in which the optic axis is uniform in all directions is called an isotropic medium.

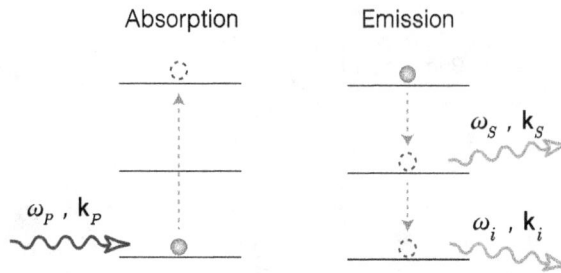

Figure 4.12. Momentum $\hbar k$ and energy $\hbar\omega$ are conserved in downconversion. Subscripts p, s, and i stand for pump, signal, and idler, respectively.

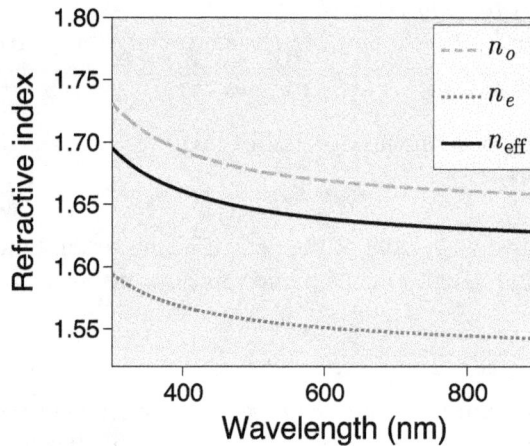

Figure 4.13. Refractive indices of BBO versus wavelength of light. The effective index n_{eff} is tuned between ordinary index n_o and extraordinary index n_e by tuning the phase-matching angle.

birefringent crystal, having *two* indices of refraction. Light is said to have ordinary polarization if it is polarized perpendicular to the optic axis of the crystal. In this case, it has an ordinary index of refraction denoted as n_o. On the other hand, for light having polarization parallel to the optic axis, there is an extraordinary index of refraction denoted as n_e.

Figure 4.13 gives the plots of the two indices of refraction of the BBO crystal whose data is provided by the crystal grower as

$$n = \left(A + \frac{B}{\lambda^2 + C} + D\lambda^2 \right)^{1/2}, \tag{4.9}$$

where the wavelength λ is in μm and the constants A, B, C, D for n_o and n_e are given by

---

[2] A medium in which the optic axis is uniform in all directions is called an isotropic medium.

- $A_o = 2.7359$
- $B_o = 0.018\,78\,\mu\,\mathrm{m}^2$
- $C_o = -0.018\,22\,\mu\,\mathrm{m}^2$
- $D_o = -0.013\,54\,\mu\,\mathrm{m}^{-2}$

- $A_e = 2.3753$
- $B_e = 0.012\,24\,\mu\,\mathrm{m}^2$
- $C_e = -0.016\,67\,\mu\,\mathrm{m}^2$
- $D_e = -0.015\,16\,\mu\,\mathrm{m}^{-2}$.

If the polarization is at an angle with respect to the optic axis, the index of refraction is modified. This effective index of refraction n_{eff} depends on the phase-matching angle, denoted as θ_m, between the propagation direction and the optic axis, and the relation is given by [35]

$$n_{\mathrm{eff}}(\theta_m) = \left(\frac{\cos^2 \theta_m}{n_o^2} + \frac{\sin^2 \theta_m}{n_e^2} \right)^{-1/2}. \tag{4.10}$$

Hence, the n_{eff} can be 'tuned' between n_e and n_o by tuning θ_m. In this way, we can get the desired index of refraction for the pump beam, which is required to satisfy equation (4.8).

In type-I phase matching, by definition, the downconverted photons have an ordinary index of refraction, and we need to tune the effective index of refraction of the pump photons so that the signal and idler beams form a laboratory angle θ_L with the pump beam. We use equation (4.9) to calculate the refractive index $n_s = n_o$ for the downconverted photons and use Snell's law $\sin \theta_L = n_s \sin \theta_s$ to obtain θ_s. Now, n_s and θ_s are inserted into equation (4.8) to determine $n_p = n_{\mathrm{eff}}(\theta_m)$ (see figure 4.13 for the tuned index of refraction), which is used to calculate the phase-matching angle θ_m using equations (4.9) and (4.10). In our case θ_m comes out to be 29.24° for $\theta_L = 3°$. Therefore, the crystals purchased were cut for a phase-matching angle of 30°. Subsequently, we use a kinematic mount to hold the mounted crystals, so fine-tuning can help us achieve the correct orientation of the crystal's optic axis by precisely tilting the BBO crystals.

4.2.2 Imaging downconverted photons

The pump laser has a wavelength of approximately 405 nm. The downconverted beams are of wavelength 810 nm, which is twice the wavelength and, hence, half the frequency of the pump beam. To ensure the separation of the two downconverted beams, phase matching is used, and the two beams are made to exit the BBO crystal, each making an angle of 3° with respect to the pump beam. The signal and idler beams emanate in opposite directions, resulting in a 6° relative angle between the two beams. In fact, only the relative angle is important. The beams need not even be horizontal and could be found in any tilted plane. Therefore, both the downconverted beams make coaxial cones that surround the pump beam, as shown in figure 4.14a.

With a proper combination of filters and lenses, these downconverted photons can be directed toward a CCD camera to form an image. Figure 4.14b shows the assembly to image the downconverted photons. The photons exit the BBO crystal at an angle of

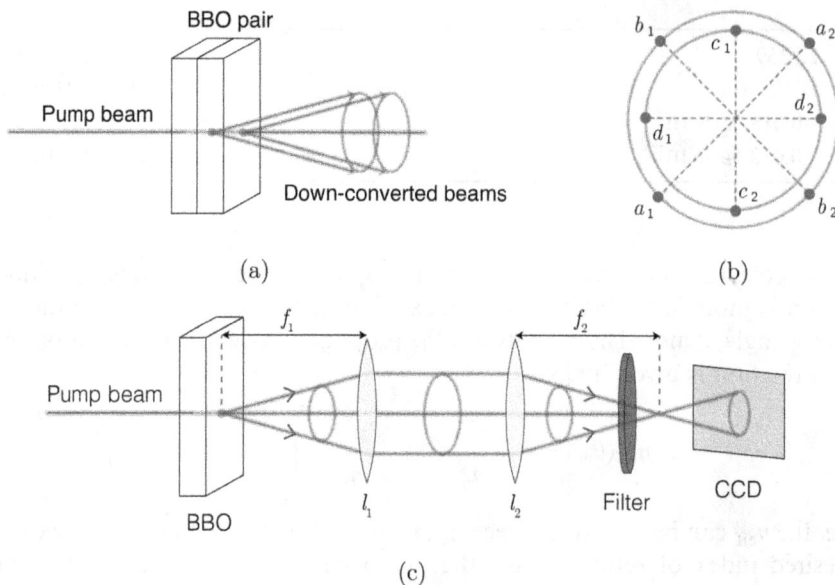

Figure 4.14. Crystal 1 downconverts horizontally polarized photons of the pump beam into vertically polarized photon pairs, and crystal 2 downconverts vertically polarized photons into horizontally polarized photon pairs. (a) Both of these crystals output the downconverted photons in the shape of a cone, while (b) shows a head-on view where correlated photons in a pair are visually illustrated by dashed lines connecting two diametrically opposed points. (c) Shows the schematic of the imaging system and illustrates its working. The expected profile of the photonic ring is also shown as it passes through the assembly of lenses l_1 and l_2.

$3°$ with respect to the pump beam. The first lens in the optical assembly makes these diverging beams into parallel beams. This can be achieved by placing the converging lens such that the focal point of the lens is at the downconversion crystal. The second lens converges these parallel beams and makes an image on the CCD, as shown in figure 4.14c. This second lens and the CCD are placed such that the CCD is further away from the lens than its focal point, and the image formed should be small enough to fit onto the CCD's surface area. Simple ray optics can be used to determine the placement of lenses. Finally, we also need to protect the CCD from the high-power pump beam. For this, we can use optical filters that do not allow 405 nm light to pass through but will permit the transmission of the 810 nm photons. Either bandpass or long-pass filters of appropriate wavelengths can be used, see figure 4.15.

The two downconversion crystals are very thin and stacked together, so in the image the two rings formed by crystals are not distinguishable and we only see a single red ring on the CCD. Now that we have seen that our source produces photons, it is time to align the entire assembly with single-photon detectors.

4.2.3 Optical alignment

If starting from scratch, one should do some back-of-the-envelope calculations and affix the major components to the optical bench for rough alignment. First, the downconversion crystal and the collection optics of one of the detectors, say A, are

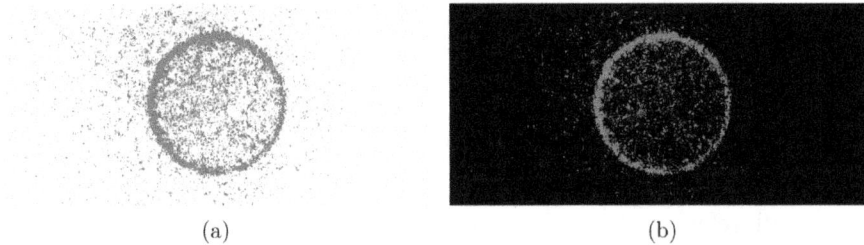

(a)   (b)

Figure 4.15. Two visual versions of a ring of downconverted photons captured on a CCD. These pictures have been cropped to focus on the ring and enhanced using standard software adjustments to improve visibility.

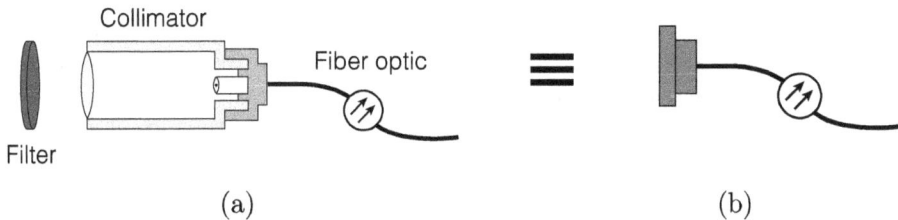

(a)   (b)

Figure 4.16. (a) The long-pass filter lets only the infrared light pass through, and the collimator directs all of the light that enters it into the core of the fiber optic cable. (b) We use the symbol, a circle with two arrows inside to designate the complete light collection package of (a).

aligned. A set of detector collection optics includes an assembly of a (long-pass) filter, a collimator, and a multimode fiber optic cable, as illustrated in figure 4.16a.

We mount each collection package on a kinematic mount, which enables horizontal and vertical tilt adjustments. The height of the lens can be modified using a post holder and a post. Details of the alignment procedure are now presented. By following these procedures, we can conveniently install the apparatus for all the experiments, and only some specific adjustments would be necessary.

For crystal alignment, the pump laser should be horizontally or vertically polarized and collimated. Some lasers, such as the one we used for our experiments, come with a polarization filter installed within the unit. In case the laser beam does not have a definite output polarization, a polarizer can be used to filter the desired polarization component. With the laser beam polarization and collimation verified, we now begin its alignment.

Using a couple of mirrors, we can align the pump beam such that it travels on the optical table or breadboard at a consistent height. Alignment can be achieved by placing the alignment ruler at multiple distances in the beam path. If the beam is aligned, there should be little to no variation in beam height, no matter how far you place the alignment ruler on the optical table. Once aligned, the pump beam is used as a reference to align all other optical objects at the same height.

First, the 405 nm HWP and the downconversion crystal are inserted in the path of the pump beam. Our laser beam is vertically polarized by default, so we set the HWP's fast axis at 0° to ensure that it does not change the polarization of the laser beam and that the BBO crystal receives vertically polarized light. To verify the action of the HWP, set $\theta = 0°$ for the $\mathbf{J}_{HWP}(\theta)$ matrix given in table 2.2 and act this on the vertically polarized input,

$$\begin{pmatrix} 1 & 0 \\ 0 & -1 \end{pmatrix}\begin{pmatrix} 0 \\ 1 \end{pmatrix} = \begin{pmatrix} 0 \\ -1 \end{pmatrix} = -\begin{pmatrix} 0 \\ 1 \end{pmatrix} \tag{4.11}$$

which means that the photons remain vertically polarized. All of the angles that we quote are with reference to the laboratory's horizontal axis.

Since the downconverted photon beams make an angle of $\pm 3°$ with reference to the pump beam, detector[3] A needs to face the downconversion crystal, roughly at $3°$ with respect to the pump beam. We can estimate the positions for detectors' placement by using simple trigonometric ratios.

A back-propagation laser is shone backwards through detector A's mount and aligned toward the downconversion crystal (figure 4.17a). We use a diode-based laser. This beam shows where detector A is pointed at. To ensure that the detector is at the appropriate height, place the alignment ruler as close to the mount as possible and check the height of the alignment laser beam. This height should be equal to the

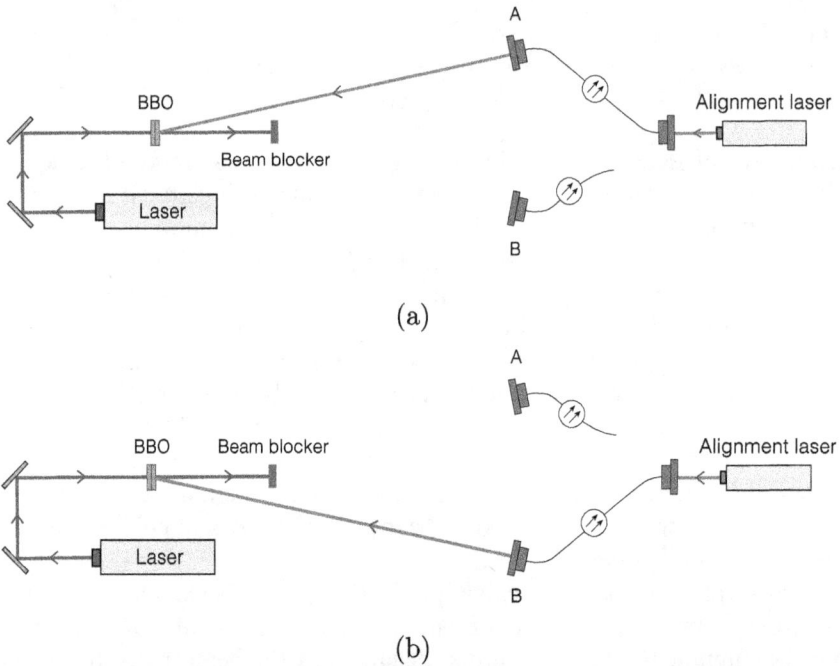

Figure 4.17. Preliminary alignment of detectors A and B. The pump beam is blocked, and a visible wavelength alignment laser is back-propagated through the collection optics of (a) detector A and (b) detector B one by one. The task is to adjust the kinematic mount of each detector to impinge the alignment beam at the center of the BBO crystals.

---

[3] For the sake of simplicity, we use the term 'detector' to collectively denote each set of the collection optics and an SPCM.

height of the pump beam. The knobs of detector A's kinematic mount are tweaked until the beam falls at the center of the downconversion crystal. Once this has been achieved, detector A has been coarsely aligned.

After the coarse alignment of detector A, we remove the alignment laser from the detector mount and then connect it to its respective SPCM. Turning the lights off and turning on the corresponding SPCM, the knobs of detector A's mount are adjusted to optimize the vertical and horizontal tilts so that the detected count rate is maximized. We can slightly slide detector A's mount sideways to optimize the beam detection angle.

Once detector A has been aligned, the next step is to align detector B such that the coincidence counts AB are maximized. Detector B is placed at a location roughly making the same angle as detector A but on the opposite side of the pump beam. The alignment laser is used to adjust the height of the detector mount and for coarse alignment (figure 4.17b). Fine alignment is then done with the SPCM connected. The knobs of detector B are adjusted to maximize the coincidence counts AB. Finally, the adjustment screws on the crystal mount are used to fine-tune the tilt angles of the crystal. After aligning both the detectors, the crystal tilt is carefully adjusted again to maximize the coincidence detection rate. It is necessary that the maximized AB counts are way above the accidental coincidence counts, which are discussed in section 4.3.7.

4.2.4 The experiment

Experiment Q1 is essentially an exercise in properly coupling the downconverted photon beams into the detectors and maximizing the coincidence counts of the two beams. Aligning the two detectors is the major requirement for this experiment. A major step in the process is aligning the collection optics, i.e., the lens, and the optical components, which bring the downconverted photons to the actual photo-detectors. This may look tedious, but once achieved this milestone becomes the launch pad for all subsequent investigations.

The schematic layout can be seen in figure 4.18, whereas a photograph of the experimental setup is shown in figure 4.19. This is an innocuously simple arrangement. Light from the pump laser, which is vertically polarized, is reflected off two

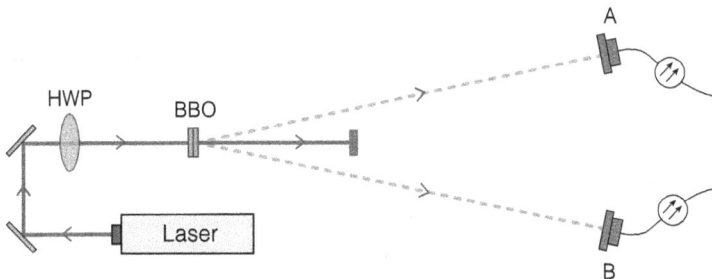

Figure 4.18. Schematic of the experimental setup for the downconversion experiment (Q1). Most of the light passes straight through and is blocked by a beam blocker. Very few photons undergo downconversion but are efficiently picked by detectors A and B. The downconverted photons (in this figure only) are shown as dashed lines.

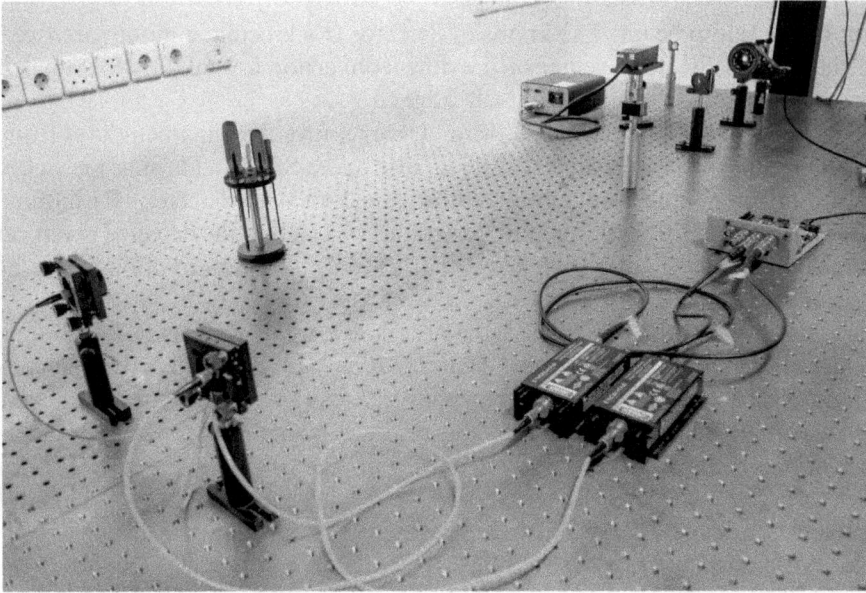

Figure 4.19. Photograph of the downconversion experiment setup for experiment Q1. The experiment is performed with the room lights turned off.

mirrors and passed through a HWP before reaching the downconversion crystals. The downconverted photons fall at the collection optics of detectors A and B. The detectors output an electric pulse for each photon that they detect. These single-photon pulses, as well as the pulses coincident on the two detectors, are counted by the CCU. These count rates are transmitted from the CCU to the computer, where they are visualized and monitored in real time.

To see if the crystals are well-aligned, the HWP is rotated. In addition, for a range of orientations of the HWP, counts registered by the detectors A and B and coincidence counts are recorded. Since we use two BBO crystals with orthogonal orientations, photons of both horizontal and vertical polarization states are down-converted. Hence, if the crystals and detectors are properly aligned, we should observe minimal or no change in coincidence counts upon rotating the HWP. We can verify this prediction using a simple line of reasoning. For example, an input polarization $|V\rangle$ creates the photon pairs $|HH\rangle$. We denote the coincidence counts in a fixed interval as κ_0. Now, rotating the input polarization by θ creates the input $\cos\theta|V\rangle + \sin\theta|H\rangle$, which results in the downconverted state $\cos\theta|HH\rangle + \sin\theta|VV\rangle$. The coincidence counts for the rotated input will become $\kappa_0(\cos^2\theta + \sin^2\theta) = \kappa_0$, and hence remain unchanged.

Figure 4.20 shows the plot of single and coincidence counts for different orientations between 0 and 180° of the HWP. The periodic change of the single and coincidence counts against the change in the HWP's orientation implies that one of the two BBO crystals is still unaligned. However, the three plots corresponding to detector A counts, detector B counts, and AB coincidence counts are all in phase, confirming the success of the SPDC process and the creation of correlated-photon pairs.

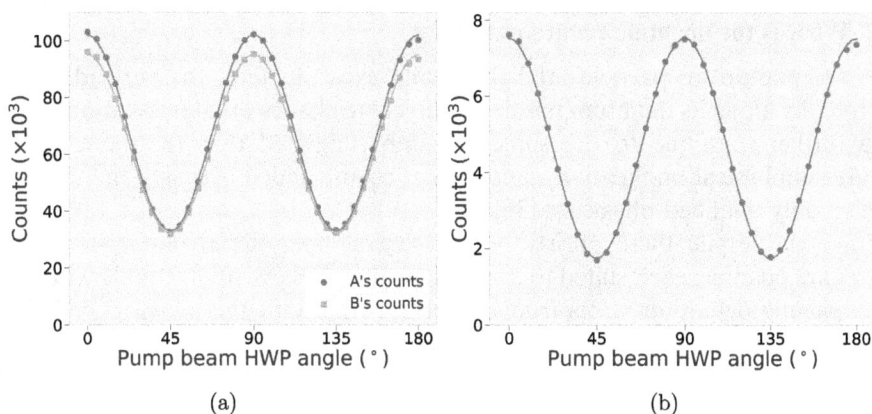

(a)   (b)

Figure 4.20. Variation of (a) single counts and (b) coincidence counts with changing orientation of the pump beam HWP before fine alignment. All the counts are in phase and show a sinusoidal variation, as depicted by the curve fits. The data shows that one of the crystals is *not* properly aligned.

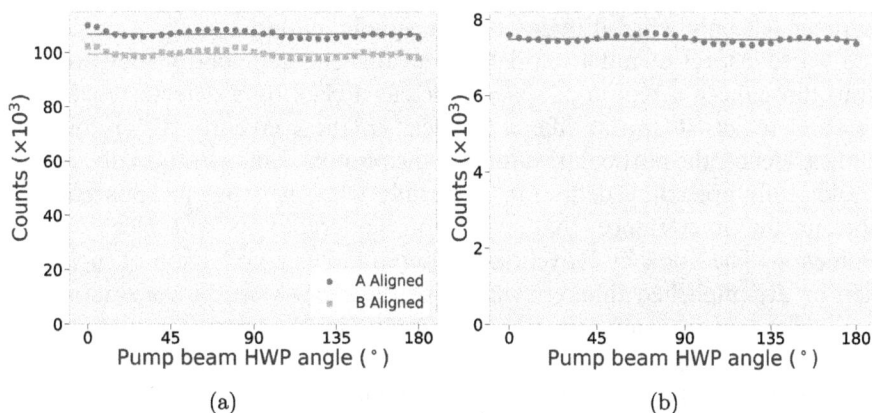

(a)   (b)

Figure 4.21. Almost no variation in (a) single counts and (b) coincidence counts with changing orientation of the pump beam HWP after the alignment has been refined. The horizontal lines show the respective mean values of the photocounts. Both of the crystals *are* now properly aligned.

If one were to refine the alignment procedure for the second downconversion crystal and then record the single and coincidence counts against the rotation of the HWP, then the single and the coincidence counts would show very little change, indicating that the BBO crystals and the detectors are well-aligned. This is illustrated in figure 4.21 and becomes an important benchmark for further experiments.

4.3 Q2: Testing the particle-like behavior of light

Experiment Q2 demonstrates the quantum nature of light and its particles, which are called photons. Before proceeding, we need to agree on what we mean by the 'quantum' nature and how it relates to photons. Let us survey the viewpoints that have been historically used in attempts to describe the photon.

4.3.1 What is the quantum nature of light?

Does the photon, a 'particle' of light, really exist? In fact, the very idea of the 'photon', let alone its quantum nature, is subject to shades of interpretation. Adding some further mystique to the subject, Lamb suggested that the word 'photon' conjures unphysical implications, and hence recommended licensing the use of this word to only qualified physicists [36].

However, despite these puristic views, several operational descriptions of the photon are possible, each suited to a specific realm. The essential trouble with an all-encompassing definition stems from the observation that this massless, relativistic *entity* has precise momentum E/c associated with an energy E (given c is the speed of light) and is normally generated from atomic level transitions. Due to the uncertainty principle, the photon is delocalized in space, and hence it is not possible to, unlike electrons and protons, regularly assign to it a wavefunction in the position space. For this reason, there is no position operator for a photon traveling in an open optical system. Therefore, some interpreters consider that a photon really exists only at the time of creation and absorption, and is fuzzy in transit. It makes its appearance felt only when it interacts with atomic matter!

The 'wave' aspect of radiation delocalizes this energetic entity (that we call the photon) throughout a region of space, whereas a click on a detector coalesces this energy and makes it appear like a 'particle' in the conventional wisdom of our semantics. Hence, the particulate nature of the photon *is* an operationally viable and psychologically appealing definition, especially when it is put in apparent contradiction to the interference that a photon undergoes as a typically wavelike phenomenon. The book by Roychoudhuri *et al* [37] is a collection of lucid articles written by accomplished thinkers who posit their views on the very nature of the photon. Alternatively, an accurate description is also found in the theory of quantum electrodynamics that presents the photon as an elementary excitation of a single mode of a quantized electromagnetic field [38]. By far, this quantum field theoretic approach is the most comprehensive picture of a photon available to us.

We now explore the quantum nature of these elusive particles. We generally seek to explain the quantum reality in opposition to a strictly classical viewpoint. Therefore, to us, quantum means nonclassical. We follow two approaches—one is based on the statistics of photodetection and the other is based on a property related to intensity measurements from a light source at two different times.

Usually, in textbooks, the quantum nature of light is introduced through the photoelectric effect. Even though Einstein's explanation of this effect is elegant and simple, there are semiclassical theories that can explain the photoelectric effect, which treat light as an electromagnetic wave and quantize the detector atoms. Such theories [39, 40] have been proposed as early as 1927. On the contrary, an experimental test to establish the existence of single photons should show the 'grainyness' of individual photons, i.e., should treat the light field quantum mechanically. These experiments should truly show the noncontinuous nature of light and the coalescing of energy into distinct shots of measurements on a detector.

Furthermore, a particle once detected should not be detected again, and a particle detected at one location should not be detected anywhere else!

Many experiments [41–43] were performed in the 1970s that demonstrated the existence of these indivisible quanta of energy, and several of them have also been adapted for the teaching laboratory [20, 33, 44]. However, the first experiments [45, 46] were based on studying the intensity correlations and were performed by Hanbury Brown and Twiss in 1956. Originally motivated by astronomical inter-ferometry, their experiments revolved around measurements of second-order (inten-sity) correlations in light. The measured correlations were higher than expected [45, 46]. Even though the semiclassical theory accurately predicted these results, this work inspired a theoretical investigation of light's statistical properties and coher-ence using purely quantum mechanical descriptions [47–49]. These studies, along with other investigative viewpoints [50–52], ultimately laid the groundwork of modern quantum optics.

In this quantum expedition, many experiments to study the statistical properties of various light sources were carried out [41, 42, 53, 54], resulting in the observation of 'antibunching' of photons. Antibunching means that photons are not found together but are spaced apart in time and are not explicable using a classical theory [43].

We now discuss photon statistics and intensity correlations—the two facets that can potentially manifest the nonclassical behavior of photons. One possible route of making progress on this front is by first exploring photon statistics from a perfectly stable laser source, which is altogether classical. If we attenuate this light beam, one might consider that at some stage we delve into the realm of *single* photons. This is, however, a naive assumption. We will describe that this chopping off the light intensity does not lay bare the photons or create quantum light. Single photons, which are the subject of this book, are therefore *not* found in a classical laser beam, no matter how dim the light is.

4.3.2 Classification based on photon statistics

The cleanest light source in the classical setting is a perfectly stable monochromatic laser beam. Suppose the power is really dim. The photons in the beam are situated at random locations. An experiment aims to receive these impinging photons and produce electronic pulses, called photocounts. The random timing of the photon leads to a random sequence of photocounts (figure 4.22).

For this beam of light, the histogram of the counts received in a time window T would be a Poissonian distribution, as discussed in section 3.7. The efficiency of the detector plays an important role in this experiment. If the detector is 100% efficient, there are no losses, and the photocount timing perfectly follows the intrinsic timing of the photons in the optical beam. An intrinsic Poissonian distribution of photons leads to a Poissionian photocount distribution. For these distributions, $\Delta N = \sqrt{\bar{N}}$. However, perfect detectors do not exist and the limited efficiency degrades the photon statistics. It can be shown [38] that if the photons have a Poissonian distribution, imperfect detections also lead to Poissonian distribution, even though the average number of counts is diminished. Furthermore, if the efficiency is much

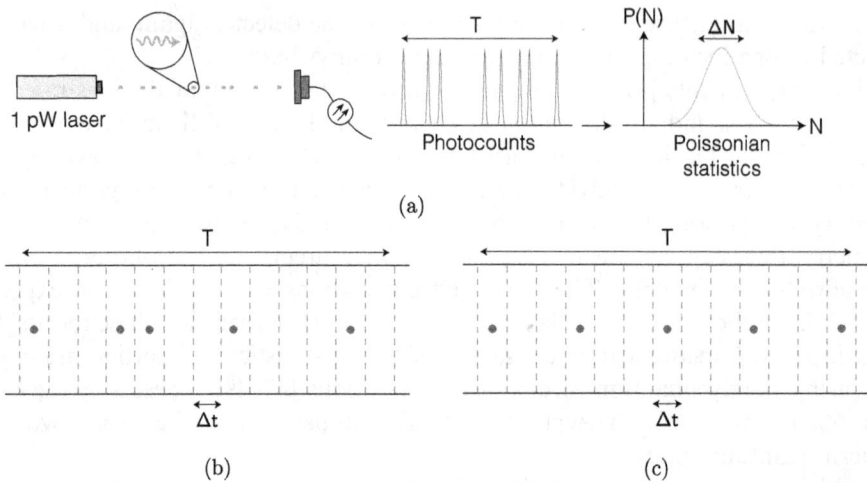

(a)

(b)   (c)

Figure 4.22. (a) A laser beam of power 1 pW produces a beam of photons that are detected. Photons are converted to photocounts by the detector. A histogram can be built showing the time of occurrence of photocounts. (b) Break the dim laser beam into subsections, each of width Δt. Some sections will have a photon, others will not. This scenario shows coherent light. (c) If the photon distribution is more organized and uniform, it is called antibunched light.

smaller than 1, the count distribution always approaches the Poissonian character, no matter what the underlying photon statistics are.

We can conceptually fragment the optical beam into smaller sections. A beam of length L takes a time L/c. Figure 4.22 shows a snapshot of this length chopped down into smaller length segments. Each bin is of duration Δt. Some bins carry a single photon, while others do not. The beam intensity is low, so the chance of two or more photons located inside a Δt bin is negligible. However, the photon distribution still remains Poissonian. Such a distribution of photons from a stable laser represents coherent light, which is the most orderly light we can obtain without invoking quantum physics. In the language of quantum optics, this is called a *coherent* beam[4].

Most sources, such as spectral lamps, light bulbs, stars, and sunlight, have a photon distribution that is many times more irregular than coherent light. The consequence is that the photon count distribution has a wider spread ΔN, i.e., $\Delta N > \bar{N}$. This light, therefore, shows a super-Poissonian distribution. Both coherent and super-Poissonian photon statistics admit perfectly classical distributions. As mentioned earlier, extremely inefficient detectors can also render intrinsic super-Poissonian photon distributions as Poissonian photocount distributions.

However, the third scenario of sub-Poissonian count statistics cannot be explained without recourse to quantum physics. In this situation, the photons are

---

[4] A coherent state $|\alpha\rangle$ is an eigenstate of the annihilation operator [38, 55], $\hat{a}|\alpha\rangle = \alpha|\alpha\rangle$ and can be expressed as a superposition of number states $|n\rangle$, $|\alpha\rangle = e^{-|\alpha|^2}\Sigma_{n=0}^{\infty} (\alpha)^n |n\rangle \sqrt{n!}$. Number states are states which carry a precise number of photons. The single-photon state is the state $|1\rangle$ in this viewpoint. Number states are also called Fock states.

uniformly spaced inside the beam as depicted in figure 4.22c, showing one bin carrying one photon interleaved with two empty bins. For a perfectly regular stream of photons, there is an occupied bin followed by k empty bins, another occupied bin followed by k empty bins and the pattern repeats itself, exhibiting no fluctuation in the photocounts, $\Delta N = 0$. Slight variations from this highly organized distribution, but still more ordered than coherent light, are also quantum. All in all, coherent photon patterns exhibiting Poissonian distribution form a convenient benchmark for light. On the one hand, more irregular patterns of photons lead to super-Poissonian statistics; and on the other hand, highly uniform photon occurrences lead to sub-Poissonian statistics. This is a trademark for nonclassical behavior. This classification is summarized in table 4.2.

If we obtain sub-Poissonian statistics, then it is a clear signature of the quantum nature of light and the existence of photons. To fully describe this light, one needs to go beyond Maxwell's approach (section 2.1). However, detecting sub-Poissonian statistics is difficult. We have hinted that inefficient detectors and optical losses that necessarily occur in any real experiment can make a sub-Poissonian photon distribution look like a Poissonian distribution, and efficient detectors are rare and very expensive! This concept is shown in figure 4.23. For this reason, only some state-of-the-art studies [43] truly depict sub-Poissonian statistics. Another approach to ascertain the quantum properties of light and the existence of photons is by

Table 4.2. Benchmarking various kinds of light. A slight subtlety, though, is to appreciate that antibunching and sub-Poissonian statistics are not necessarily from the same underlying reason. These manifestations often appear together, but they are independent properties. This subtlety is explored further in [56].

$g^{(2)}(0)$	Photon statistics	Property of light	Examples
>1	Super-Poissonian	Bunched	Starlight, sunlight, incandescent bulb.
1	Poissonian	Coherent	An ultra-stable monochromatic laser.
<1	Sub-Poissonian	Antibunched	Single-photon states (the topic of this book).

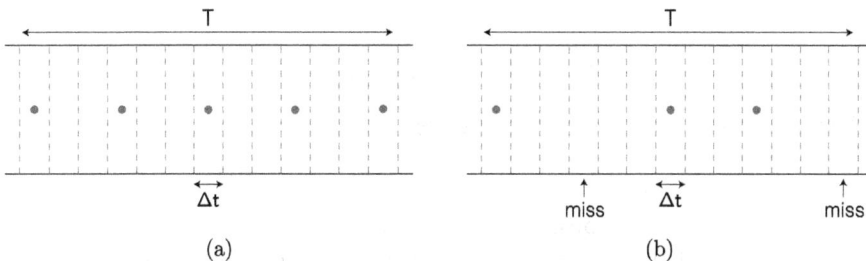

Figure 4.23. Due to inefficient detections, some of the bins that had photons were not detected. Making a sub-Poissonian photon distribution (a) seem like a Poissonian distribution (b).

observing intensity correlations in the style of Brown and Twiss. The quantifiable metric that helps to explore this is the degree of second-order coherence $g^{(2)}(0)$.

4.3.3 Classification based on intensity (anti)correlations

In this vein, in 1986 Grangier *et al* performed an elegant and conceptually simple experiment [28, 57]. The idea was to study correlations between photodetections at the transmission and reflection sides of an ordinary 50:50 beam splitter (BS). If the beam consisted of individual photons, then just one of the detectors (either transmission or reflection) would register a count at any given instance, and there would be no coincident detections. This experiment showed that only a single-photon state could account for the field incident on the BS [28]. However, the key challenge in such an experiment is *creating* a true single-photon state, which a dim-intensity coherent beam does not qualify to be.

In this book, we describe an updated version of the Grangier experiment, closely following the work described in the references [3, 25, 28]. We use a downconversion-based heralded photon source to ensure the incidence of single photons on the BS, as described in experiment Q1. Photons are produced in pairs through the SPDC process. A click on the signal must coincide with a click on the idler if these photons are truly a correlated pair. One click for one photon! As the measurements of signal beam photons are conditioned on the detection of the idler beam on a certain detector, the signal beam effectively becomes a single-photon state.

We can quantify the correlation in the photodetection process by a parameter known as the second-order correlation function [55], $g^{(2)}(\tau)$. Correlation functions quantify the association between a property of two beams of light or a single beam of light at different time or space points. A first-order (temporal) correlation function measures the correlation between the electric field strengths in time and is given by

$$g^{(1)}(\tau) = \frac{\langle E^*(t)E(t + \tau)\rangle}{\langle |E(t)|^2\rangle}, \tag{4.12}$$

where $\langle \cdots \rangle$ represents an average over time. The anticorrelation experiments deal with intensities instead of electric fields. Since intensity is proportional to the square of the electric fields, the intensity correlation function is of *second-order* and is given by

$$g^{(2)}(\tau) = \frac{\langle I_2(t)I_1(t + \tau)\rangle}{\langle I_2(t)\rangle\langle I_1(t + \tau)\rangle}. \tag{4.13}$$

For the three kinds of light, the second-order correlation function takes up values in the ranges specified in table 4.2. Therefore, if our measurements result in a value of $g^{(2)}(0)$ less than 1, then we can pronounce that classical theories are unable to explain the optical field. This will constitute an experimental proof of the granular nature and the nonclassical disposition of light. The quantum state that would show a maximal violation of this inequality, $g^{(2)}(0) = 0$, is *the* single-photon state. This value of $g^{(2)}(0) = 0$ implies anticoincidence between the two output ports of the BS.

A photon can only be detected on one of the two output channels. This experiment is, of course, conditioned on the gating photon received on the idler beam. We discuss the meaning of $g^{(2)}(0) = 0$ in the following section.

4.3.4 Predicting the degree of second-order coherence

In this section, we model the photodetection process and present the classical, semiclassical, as well as quantum predictions for $g^{(2)}(0)$, all in the same order. The classical view applies when light is treated as a classical wave, and the photo-detection is also classical, modeled by a BS separating the light flux into two parts. The semiclassical approach makes the distinction that it models the detection by the ejection of electrons when light strikes on the detector. The ejected photoelectrons give the photocounts and this happens discretely with a certain efficiency while light is still treated as a classical wave. Finally, the quantum view also treats the light beam as a collection of discretely spaced photons.

4.3.4.1 The classical view

When we talk about classical fields, we refer to electromagnetic waves whose behavior are completely encompassed by Maxwell's equations (see equations (2.1)–(2.4)). Referring to figure 4.24, consider an optical field directed toward a BS. Some part of this incident beam is transmitted and falls on detector B, and the other part is reflected and falls on detector B'.

Let $I_I(t)$ denote the intensity of the input field and let $I_B(t)$ and $I_{B'}(t)$ represent the respective intensities of the beams falling on the detectors. Then, to gauge the correlations between the detector intensities $I_B(t)$ and $I_{B'}(t)$, we use the second-order correlation function. This is denoted as $g^{(2)}_{B, B'}(\tau)$ and expressed as [3, 55]:

$$g^{(2)}_{B, B'}(\tau) = \frac{\langle I_{B'}(t)I_B(t + \tau)\rangle}{\langle I_{B'}(t)\rangle\langle I_B(t + \tau)\rangle}, \qquad (4.14)$$

where τ represents the time delay between the two intensity measurements.

We are interested in simultaneous intensity correlations, i.e., we would like to study $g^{(2)}_{B, B'}(\tau)$ at $\tau = 0$. Suppose we represent the transmission coefficient of the splitter as T and the reflection coefficient as $R = 1 - T$. In that case, it is straightforward to see that the transmitted and reflected intensities are given by

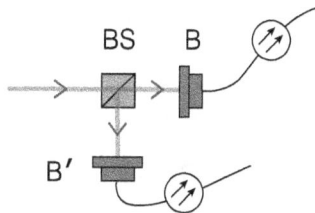

Figure 4.24. The 50:50 BS separates the incident light into two output channels, one received on detector B and the other received on detector B'.

$I_B(t) = TI(t)$ and $I_{B'}(t) = RI(t)$, where $I(t)$ is the input light intensity, purely classical. Inserting these expressions into equation (4.14) gives

$$g^{(2)}(0) = g^{(2)}_{B, B'}(0)$$

$$= \frac{\langle [I(t)]^2 \rangle}{\langle (t) \rangle^2}. \tag{4.15}$$

Let us first look at the numerator $\langle I^2(t) \rangle$. Since this is a time average, we can write

$$\langle I^2(t) \rangle = \frac{I^2(t_1) + I^2(t_2) + ... + I^2(t_s)}{s}$$

$$= \frac{\sum\limits_{i=1}^{s} I^2(t_i)}{s}, \tag{4.16}$$

where s points are taken to compose the average. Likewise, for the denominator, we have

$$\langle I(t) \rangle^2 = \frac{(I(t_1) + I(t_2) + ... + I(t_s))^2}{s^2}$$

$$= \left(\frac{\sum\limits_{i=1}^{s} I(t_i)}{s} \right)^2. \tag{4.17}$$

The square of the sum of these terms can also be written as the pairwise product of intensities, therefore,

$$\langle I(t) \rangle^2 = \frac{\sum\limits_{j=1}^{s}\sum\limits_{i=1}^{s} I(t_i)I(t_j)}{s^2}. \tag{4.18}$$

Now we know that $I^2(t_1) + I^2(t_2) \geqslant I(t_1)I(t_2) + I(t_2)I(t_1)$ (simply following from $x_1^2 + x_2^2 \geqslant 2x_1 x_2$), and therefore each of these terms in the pairwise product must be smaller than the corresponding sum of squares. Spelling this out for the entire sequence, we have

$$\sum\limits_{j=1}^{s}\sum\limits_{i=1}^{s} I(t_i)I(t_j) \leqslant s \left(\sum\limits_{i=1}^{s} I^2(t_i) \right). \tag{4.19}$$

Comparing equations (4.18) and (4.19), we are able to show that

$$\langle I(t) \rangle^2 \leqslant \frac{\sum\limits_{i=1}^{s} I_i^2(t)}{s} = \langle I^2(t) \rangle,$$

clearly indicating that

$$g^{(2)}(0) \geqslant 1. \tag{4.20}$$

If we have perfectly stable light with no fluctuations ($I(t) = I_0$), then we can, using $\langle I^2(t) \rangle = \langle I_0^2 \rangle = I_0^2$, also deduce that for this coherent beam,

$$g^{(2)}(0) = 1, \tag{4.21}$$

which is a special case of perfectly time-invariant intensity.

The equality with one is obtained if there are no fluctuations in the input intensity. In contrast, if we have a fluctuating input field intensity, $g^{(2)}(0)$ is found to be greater than 1. The best approximation for a stable input source that is monochromatic, an ultra-stable laser light, achieves $g^{(2)}(0) = 1$. Another example of classical light is the so-called 'chaotic' light [55], which possesses $g^{(2)}(0) = 2$. An example of this is light received from a vapor lamp. In short, classical light is bound to yield a value of $g^{(2)}(0)$ greater than or equal to 1.0 and if a smaller value is attained, then it will constitute an indication of quantum light.

In single-photon experiments, we do not directly measure the intensity; instead, we measure the output current from a photodetector. Hence, we now discuss the process of photodetection. We must not lose sight of the fact that detection statistics result from the bunching and antibunching properties of photons inside the beam, as well as, given the detector efficiency, the statistics of how frequently the detectors output photocurrent signals. For $g^{(2)}(0) > 1$, there is a high likelihood of observing coincidences on the output ports of the BS. These photons are more likely to be produced together than apart. This property is called photon bunching [38].

4.3.4.2 The semiclassical view with two detectors

We now need a practical strategy to measure $g^{(2)}(0)$ defined in equation (4.15). This requires expressing the formula in terms of photon counts. According to the semi-classical theory of photoelectric detection [55], photoelectrons are randomly generated due to incident electromagnetic waves. Therefore, technically speaking, the one-photon one-click rule does not apply. Using a photodetector (say detector B), the probability of registering one photocount within a short interval represented by Δt is given by

$$P_B = \eta_B \langle I_B(t) \rangle \Delta t, \tag{4.22}$$

where $\langle I_B(t) \rangle$ represents the time-averaged intensity of the field incident on the detector and η_B denotes the photodetection efficiency. Likewise, the *joint* probability of registering a photocount (in a short window Δt) at detector B', followed by another photocount at detector B after time τ (also within the identical time window Δt), is given by

$$P_{BB'}(\tau) = \eta_B \eta_{B'} \langle I_B(t + \tau) I_{B'}(t) \rangle (\Delta t)^2. \tag{4.23}$$

Interestingly, $P_{BB'}(\tau)$ contains two kinds of paired detections: B clicking τ seconds after B', as well as B clicking τ seconds before B'. From equations (4.14), (4.22), and (4.23), we obtain

$$g_{BB'}^{(2)}(\tau) = \frac{P_{BB'}(\tau)}{P_B P_{B'}}. \tag{4.24}$$

Therefore, the degree of second-order coherence can be determined by measuring the probability of single and coincident photodetections at detectors B and B', separated in time by τ. For $\tau = 0$, we have

$$g_{BB'}^{(2)}(0) = \frac{P_{BB'}(0)}{P_B P_{B'}}. \tag{4.25}$$

This formula gives us a practical way to determine the second-order g function. The only task that remains is to estimate the probabilities.

The probability of registering a photodetection at detector B within a small interval Δt can be *estimated* by multiplying the average photodetection rate R_B with Δt. To compute the average photodetection rate for detector B, we divide the number of photodetections N_B by the total time T over which we count, where T is sometimes called the integration time. We can determine the probabilities for individual detections at detector B' ($N_{B'}$) and BB' coincidence detections ($N_{BB'}$) in a similar fashion:

$$P_B = \left(\frac{N_B}{T}\right)\Delta t = R_B \Delta t, \tag{4.26a}$$

$$P_{B'} = \left(\frac{N_{B'}}{T}\right)\Delta t = R_{B'} \Delta t, \tag{4.26b}$$

$$P_{BB'} = \left(\frac{N_{BB'}}{T}\right)\Delta t = R_{BB'} \Delta t. \tag{4.26c}$$

In our case, the variable Δt corresponds to the coincidence time window of the CCU. Substituting equations (4.26) into equation (4.25), we obtain

$$g_{BB'}^{(2)}(0) = \frac{N_{BB'}}{N_B N_{B'}}\left(\frac{T}{\Delta t}\right). \tag{4.27}$$

This is the practical formula that we have been seeking. It quantifies correlations between the output ports of the BS. For classical light, we expect this function to be greater than one. It will also be greater than one if we do not consider the heralding process, without which we do not create single photons. However, the unheralded results can become a useful control for comparing with the heralded scheme that we describe next.

4.3.4.3 The quantum view with the heralded scheme and three detectors
In a three-detector experiment, we still measure $g^{(2)}(0)$ for a signal beam incident on a BS, but the difference is that the photodetection of this beam is now *conditioned* on the photodetection of another idler beam. As mentioned earlier, the conditioning is

crucial—photodetection of the idler beam ensures the single-photon (quantum) state of the signal beam. Let us now discuss the three-detector experiment.

Consider the measurement of $g^{(2)}(0)$ using a three-detector setup shown in figure 4.25. For this setup, analogous to equation (4.25), we obtain the following expression for the second-order correlation function:

$$g^{(2)}_{ABB'}(0) = \frac{P_{ABB'}(0)}{P_{AB}(0)P_{AB'}(0)}, \tag{4.28}$$

where $P_{ABB'}(0)$ represents the probability of a three-fold coincidence event involving all the three detectors, and $P_{AB}(0)$ and $P_{AB'}(0)$ represent the probabilities of two-fold coincidence events at the detectors A and B; A and B'. A click at detector A signals the presence of that coveted single photon in the other beam, which is subsequently projected onto either B or B', *but not both.* If we were to discount clicks on A, then we would not observe this mutual exclusivity between clicks on B and B' because the unheralded photons are not guaranteed single-photon states. Therefore, this experiment is set to verify this particulate prediction and ascertain the grainyness of the photon.

We want to monitor only the events in which detector A is triggered. Hence, we can take the number of photodetections at this detector and use it as the total number of trials. Representing this number as N_A, we use it to normalize the photodetection probabilities as follows:

$$P_{AB}(0) = \frac{N_{AB}}{N_A}, \tag{4.29a}$$

$$P_{AB'}(0) = \frac{N_{AB'}}{N_A}, \tag{4.29b}$$

$$P_{ABB'}(0) = \frac{N_{ABB'}}{N_A}. \tag{4.29c}$$

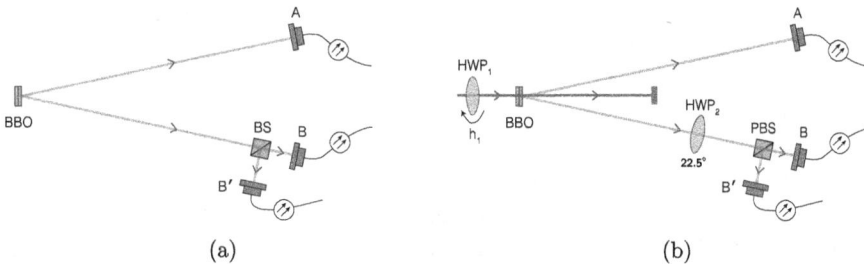

(a)   (b)

Figure 4.25. The BBO source produces pairs of photons, one of which goes to detector A and the other goes to the (a) BS, with two detectors B and B' placed at the transmission and reflection outputs of the polarizing BS. Detection of a photon at detector A projects the other photon into a single-photon state. In (b), the BS is replaced by a polarizing BS, which necessitates the use of HWPs to adjust the polarization of photons. See text for details.

Plugging these probabilities into equation (4.28) yields the quantum prediction for the degree of second-order coherence,

$$g_{ABB'}^{(2)}(0) = \frac{N_A N_{ABB'}}{N_{AB} N_{AB'}}. \tag{4.30}$$

If an individual photon is incident on the BS shown in figure 4.25, we should obtain $N_{ABB'} = 0$ in equation (4.30) yielding the prediction $g^{(2)}(0) = 0$ for non-classical light. In the case that the photodetections being monitored at detectors B and B' are totally uncorrelated, we obtain (in the two-detector scheme) $g_{BB'}^{(2)}(0) = 1$, corresponding to a coherent beam of photons. The value can be even larger than 1 for chaotic light.

This argument can also be understood intuitively. The function of a BS is to simply divide an optical field into two daughter fields. These resultant fields may either swing together or may not swing at all. In the former case, there is a positive correlation in the fluctuations of the field, which is called 'bunching'. In the latter case, there are no fluctuations to begin with, and hence no meaning of correlations.

The third possibility is an increase in the intensity of one of the daughter fields while, at the same time, there is a decrease in the intensity of the other daughter field. Clearly, this is infeasible from a classical standpoint. Such a situation would imply anticorrelations and is termed the 'antibunching' of photons, wherein the creation of a photon precludes the coexistence of another photon. Photons that are spaced apart are favored as compared to photons that coexist in bunches—hence the name 'antibunching' [38].

We would like to point to a subtlety involved in this experiment. To ensure a measurement approaching $g^{(2)}(0) = 0$, we are required to ensure single-photon incidence on the polarizing beam splitter (PBS). Using our SPDC source, this requirement is achieved by conditioning the signal photon detection to an idler photon seen on detector A. Therefore, for such a light source, the requirement of a three-detector setup is crucial to obtain $g^{(2)}(0) = 0$. If we employ the downconversion source and do not perform the conditioning of detection, we will not achieve single-photon incidence on the BS. Measurements with this setup will consequently result in $g^{(2)}(0) \geqslant 1$. In our experiments, we use the same setup for two-detector and three-detector $g^{(2)}(0)$ measurements but look at different sets of single and coincidence counts, as required by equations (4.27) and (4.30), respectively, for comparing the two schemes.

Finally, while performing an experiment in the real laboratory setting, we do not expect to achieve an ideal measurement that categorically yields $g^{(2)}(0) = 0$. This is due to the presence of accidental coincidences and photons with imperfect quantum purity [58, 59]. We can predict these accidental coincidences using the coincidence time window Δt, the integration time T, and the average values of the single detector counts. Using these predicted accidental coincidence detections, an uncertainty value can be quoted alongside the expected $g^{(2)}(0)$. The calculation of accidental coincidences is outlined in section 4.3.7.

4.3.5 Preparing the experiment

In experiment Q1, we examined the behavior of an SPDC source. In the alignment we attempted to maximize the coincidence detection rate between two detectors, namely A and B. Experiment Q2 builds up on the same setup, where detectors A and B have already been aligned. Figure 4.25a shows a proposed setup, whereby the signal photons fall on a BS and are led to two possible paths B and B'.

An alternative scenario is shown in figure 4.25b wherein the BS is replaced by a PBS. The PBS has two output ports: one corresponding to $|V\rangle$, which leads to the B detector, and one to $|H\rangle$ making its way to the B' detector. Some additional modifications to this scheme are, however, needed. For example, a $|V\rangle$ photon incident on the PBS will *always* emerge on the transmitted channel, and hence there will be perfect anticorrelation. However, this is unintended and does not emanate from the nonclassical properties of the photon. On the other hand, a $|D\rangle = (|H\rangle + |V\rangle)/\sqrt{2}$ photon will have equal potentiality to emerge on either of the two output detectors in an unbiased fashion. This means that any anticoincidence on B and B' will then truly represent a property of the photon, unbiased from its polarization degree of freedom. This adjustment to the polarization incident on the PBS can be achieved with the help of two HWPs, HWP_1 and HWP_2, in the pump and signal beams, respectively, so that the likelihood of emerging from the two output channels of the PBS is equal. Subsequently, it is predicted that in this case of polarization tagging, perfect anticorrelation between the B and B' clicks will also be observed for single-photon states. Our experiment Q2 implements the very scheme presented in figure 4.25b. In the first step, we align the components.

4.3.5.1 Aligning the detectors
For Q2, the alignment process is described in figure 4.26. This stage of alignment does not need the pump beam. The major task in this experiment is to insert a PBS in the signal beam path and then align the detector B'. As soon as this is achieved, we can measure $g^{(2)}(0)$. We already have the detector B aligned in the signal beam path from experiment Q1, and we will use it as a reference starting point. The PBS is inserted just about a distance of 4–5 inches from the detector B, and an alignment laser is attached with the detector B mount. This laser beam will appear as a spot on

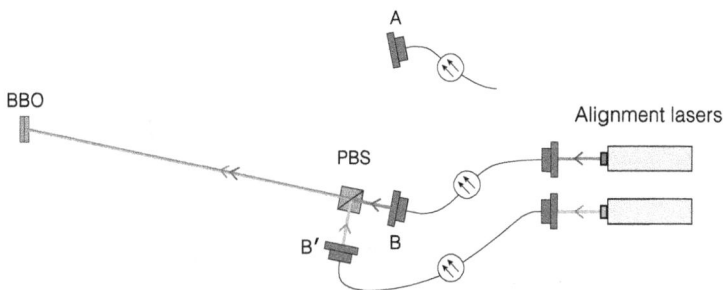

Figure 4.26. Coarse alignment of detectors B and B' with a PBS placed in the signal path. Visible wavelength, unpolarized alignment lasers are back-propagated through the collection optics of detectors B and B'.

the BS, and a portion of it will transmit and fall on the BBO crystal. Alignment lasers are unpolarized.

Next, we aim to align detector B' such that both detectors B' and B receive light from the same downconverted beam. The collection optics of the detector B' are placed in such a way that they receive the light from the source that is reflected by the BS[5]. It is ensured that both detectors B' and B are equidistant from the BS and detector B' is also at the same height as the rest of the components.

After placing the detector B' at an appropriate position, an alignment laser (preferably of a different color) is back-propagated through its mount, as shown in figure 4.26. We need to ensure that the spot that this laser makes on the BS overlaps with the spot made by the laser coming from detector B. Furthermore, we also need to ensure that the light from detector B' being reflected by the PBS also falls onto the BBO crystal. To ensure that both of these conditions are met, the tilts of both the detector mount and the PBS need to be adjusted. Once this much is achieved, we can consider the coarse alignment complete.

4.3.5.2 Creating the desired polarization
The next step is to create the desired polarization state to be fed into the PBS in the signal path. This is crucial so that the steering of a photon to either of the output channels is unbiased with respect to the polarization, allowing one to deduce genuine coincidences, or lack thereof. This polarization adjustment is achieved through two HWPs: HWP_1 in the pump beam and HWP_2 in the signal beam. Suppose their orientation angles are h_1 and h_2 respectively. The complete system will follow the scheme shown in figure 4.27 and photographed in figure 4.29.

We set $h_1 = 0$ and vary h_2 to maximize the counts B' and the coincidences AB'. This will set the HWP_2 to $h_2 = 0$. Now turn HWP_2 to approximately $h_2 = 45°$. Ideally, this setting of h_2 should maximize B and AB. Given the two endpoints of the swing, turn the HWP_2 *back* by 22.5° to achieve $h_2 = 22.5°$. This setting should now equalize the counts AB and AB'. After all, what has this achieved? The pump polarization is vertical (given $h_1 = 0$), leading to the generation of the

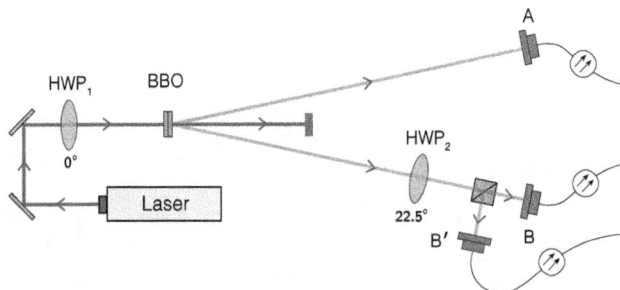

Figure 4.27. Schematic diagram for experiment Q2.

---

[5] Beams splitters usually have arrows marking the reflection and transmission side of the BS.

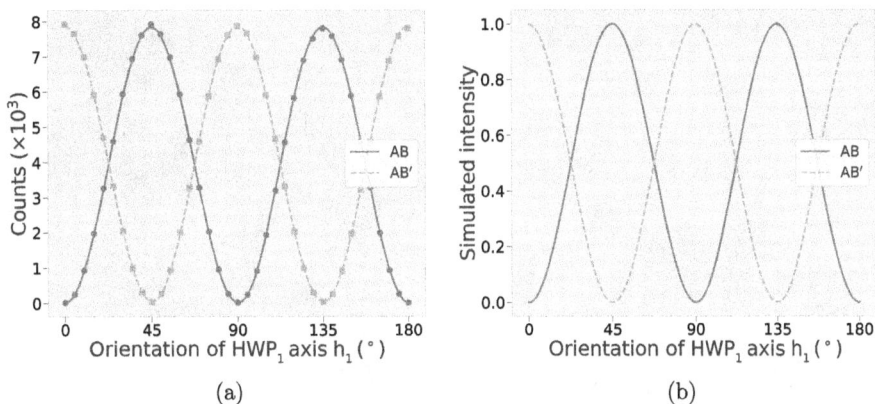

(a) (b)

Figure 4.28. Variation of AB and AB' coincidence counts with respect to change in orientation of the pump beam HWP (HWP_1). (a) Measured coincidence counts with curve fits $7824\cos^2(2h_1) + 18$ and $7814\sin^2(2h_1) + 48$, corresponding to AB and AB' respectively. (b) Simulated coincidence counts where AB' represents the photon pairs predicted in the state $|HH\rangle$; and AB represents the photon pairs predicted in the state $|VV\rangle$.

downconverted photons $|HH\rangle$. The setting $h_2 = 22.5°$ creates $\left(\frac{|H\rangle + |V\rangle}{\sqrt{2}}\right)$, which ensures that the photons falling onto the PBS are directed to the two detectors B and B' in a completely balanced fashion, allowing us to infer genuine correlations for the measurement of $g^{(2)}(0)$.

It is recommended that there should be very little difference between the maximal values for AB and AB' coincidence detections. If the counts are drastically different, alignment may need to be improved or redone. The optic fibers for one of the B and B' detectors may also need cleaning, or it might be the case that the fiber-coupling lens for one of the detectors may have a better alignment than that of the other. This would then need readjustment.

We can perform another final check, if we wish, to see if the assembled system is working accurately. For this part, remove HWP_2 and rotate HWP_1 inside the pump beam (varying the angle h_1). Changing h_1 creates a pump state $\cos(2h_1)|V\rangle - \sin(2h_1)|H\rangle$ which, after passing through the BBO pair, becomes $\cos(2h_1)|HH\rangle - \sin(2h_1)|VV\rangle$ upon downconversion. Using equations (5.22) and (3.23), we see that the probability that detector B receives a photon is $\sin^2(2h_1)$ and the probability of receiving a photon on detector B' is $\cos^2(2h_1)$. Simulated and experimentally inferred results are drawn in figure 4.28, showing the excellent agreement between prediction and experiment. At this stage, all is set for experiment Q2, and we can reinsert HWP_2 into the signal path at the orientation $h_2 = 22.5°$.

4.3.6 Experimental results

Once a satisfactory level of alignment is achieved, preparatory tests described in the preceding storyline have been performed, and the wave plates are set at $h_1 = 0$ and $h_2 = 22.5°$, we can record counts registered at detectors A, B, and B', and the coincidence counts for AB, AB', BB', and ABB'. Then, using the expressions in

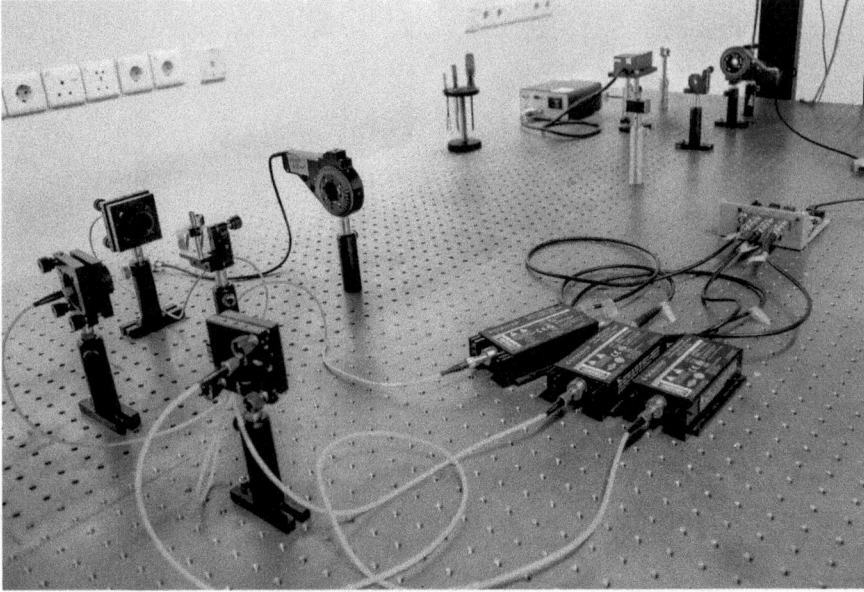

Figure 4.29. Photograph of the three-detector $g^{(2)}(0)$ measurement setup.

Table 4.3. Table showing typical datasets obtained from experiment Q2 for different coincidence windows. All these datasets comprise of 20 time windows each of 1 s and the means are quoted. All rates are quoted in counts per second (cps). The numbers in parentheses show the uncertainty in the least significant digits in the quoted mean values. The uncertainties are determined by the standard error formula. The column labeled confidence indicates the number of standard errors by which the classical result is violated and is a measure of the degree of confidence in nonclassicality.

Δt (ns)	R_A	R_{AB}	$R_{AB'}$	$R_{ABB'}$	$g^{(2)}_{ABB'}(0)$	$g^{(2)}_{ABB'}(0)^{(acc)}$	Confidence (σ)
40	38 447(76)	504(6)	519(5)	0.70(28)	0.107(41)	0.119	22
20	103 892(393)	3907(66)	4037(58)	4(2)	0.0211(3)	0.057	343
3	128 767(337)	4254(62)	4360(52)	6(2)	0.042(3)	0.011	310

equations (4.27) and (4.30) for the two-detector and three-detector configurations, one can compute the second-order correlation functions. The key results derived from a sample of three distinct experimental sessions are summarized in table 4.3.

The value of $g^{(2)}_{ABB'}(0)$ for each dataset is calculated using equation (4.30). All of these results and the many more available on our website confirm that the second-order correlation function is less than one and very close to zero, as expected of single-photon quantum states. Furthermore, the values are comparable to the accidental coincidences, which are described in the next subsection. Each of these results indicates a marked departure from classical predictions. The quantum violation is aptly described by the number of standard errors away from the

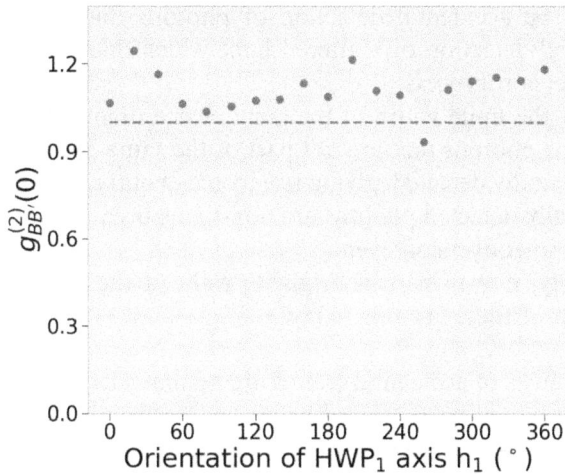

Figure 4.30. Two-detector (classical) degree of second-order coherence $g_{BB'}^{(2)}(0)$ for various orientations of the HWP (h_1). The dashed line corresponds to $g_{BB'}^{(2)}(0) = 1$.

classical value, computed as $(1 - g_{ABB'}^{(2)}(0))/\Delta g$, where Δg is the uncertainty in $g_{ABB'}^{(2)}(0)$, calculated from the count rates and presented in table 4.3. The results clearly indicate the quantum nature and antibunching of the single photons.

It is also possible to estimate the two-detector correlation function $g_{BB'}^{(2)}(0)$, which would typify classical light. However, our FPGA code is not directly programmed to measure BB' coincidences (as can be seen in the sequence of information sent inside a communication cycle, shown in figure 4.8). Therefore, for the two-detector measurement, the output cabling from the SPCM's is rerouted so that the counts N_B, N_B' and $N_{BB'}$ become available. With this information, equation (4.30) can be employed to estimate the correlation function. In one run of experiments, we varied h_1 in steps of 20° within the 360° span and obtained the correlation function, which remained very close to one, as expected for classical light. From a set of 20 measurements shown in figure 4.30, we obtained $g_{BB'}^{(2)}(0) = 1.110 \pm 0.015$, which is close to the classical expectation of one.

4.3.7 Accidental coincidence counts

One of the questions that may arise is that the quantum picture predicts a perfect anticorrelation, which implies $g^{(2)}(0) = 0$ precisely. So, even though $g^{(2)}(0) < 1$ is in accordance with the granular existence of photons, we are concerned about why a value greater than zero is detected at all. This means that in equation (4.29), the triple coincidence $N_{ABB'}$ must be zero. In fact, this is nonzero due to accidental coincidences. We must estimate how frequent these accidentals are. Accidental counts contribute to false positives and have the potential to overestimate measured counts. There could be several sources for these coincidences:

1. There will be accidental detection of photons due to dark counts of the detector and detection of scattered light which does not originate from the downconversion process.
2. Because of the finite width of the coincidence time window and misaligned optics, some photons that are not part of the same downconversion pair will be coincidentally detected, giving rise to accidental coincidence counts. These coincidentally detected photons are non-twins because they do not belong to a single downconversion event.
3. Occasionally, non-twins can originate right at the BBO crystal due to the finite width of these crystals.

All in all, these sources of accidental counts are assumed to be random, but they can be quantifiably estimated, as discussed below.

Let us consider two-fold coincidences on detectors A and B. The probability for such an accidental coincidence is proportional to the probabilities for count registration on the two detectors,

$$P_{AB}^{(acc)} = P_A P_B. \tag{4.31}$$

This assumes that clicks on A and B are due to random events that are independent and uncoupled. Now, using the probability estimates in equation (4.25),

$$P_A = \left(\frac{N_A}{T}\right)\Delta t = \frac{N_A}{N_P} \tag{4.32}$$

$$P_B = \left(\frac{N_B}{T}\right)\Delta t = \frac{N_B}{N_P} \tag{4.33}$$

$$P_{AB} = \left(\frac{N_{AB}}{T}\right)\Delta t = \frac{N_{AB}}{N_P} \tag{4.34}$$

where $N_P = (T/\Delta t)$ is the total number of *possible* counts, the accidental probability becomes

$$
\begin{aligned}
P_{AB}^{(acc)} &= \frac{N_A}{N_P}\frac{N_B}{N_P} \\
\frac{N_{AB}^{(acc)}}{N_P} &= \frac{N_A}{N_P}\frac{N_B}{N_P} \\
N_{AB}^{(acc)} &= \frac{N_A N_B}{N_P} = \frac{N_A N_B}{T}\Delta t,
\end{aligned}
\tag{4.35}
$$

which allows us to estimate the number of accidental two-fold coincidences originating from random, independent events. Using $R_i = N_i/T$ for $i = A, B, AB$, one can express this quantity in terms of rates as well,

$$R_{AB}^{(acc)} = R_A R_B \Delta t \tag{4.36}$$

which is identical to the expression for two-fold accidental rates given in [60, 61]. Similarly, genuine three-fold coincidences between three independent, purely random events are given by [61]

$$R_{ABB'}^{(acc)} = 3R_A R_B R_B'(\Delta t)^2. \tag{4.37}$$

Clearly, if $\Delta t \ll T$, then this term is negligible and can be ignored. All in all, the measured coincidences will have a true value and an accidental component originating from non-twins and the background,

$$R_{AB} = R_{AB}^{(true)} + R_{AB}^{(acc)}. \tag{4.38}$$

We do not have (or have a negligible) number of genuine three-fold coincidences (equation (4.37)), but it is possible to have a genuine two-fold coincidence synchronized with a random signal count on the third detector. This leads to a term $P_{ABB'}^{(acc)}$, which is used in the calculation of the three-fold second-order correlation function $g^{(2)}(0)$, given by equation (4.28) (in terms of probabilities) or equation (4.30) (in terms of counts). In this likely scenario, an accidental $N_{ABB'}$ may occur in two ways. First is a genuine AB coincidence, which arrives at the same time (within a coincidence window) as an independent, random click on B'. The second is the converse, an AB' coincidence coexisting with a B click. These scenarios allow us to estimate the probability $P_{ABB'}^{(acc)}$ as

$$P_{ABB'}^{(acc)} = P_{AB} P_{B'} + P_{AB'} P_B. \tag{4.39}$$

In the three-fold coincidence experiment, the total number of accidental counts is given by the click on detector A because these are all that matter. Therefore, equation (4.34) can be written for the three-fold case as,

$$P_{AB} = \frac{N_{AB}}{N_A}, \quad \text{and} \tag{4.40}$$

$$P_{AB'} = \frac{N_{AB'}}{N_A}, \tag{4.41}$$

whereas, the signals P_B and $P_{B'}$ can be written identically to equations (4.32) and (4.33) as

$$P_B = \frac{N_B}{N_P} = \frac{N_B}{T}\Delta t = R_B \Delta t, \quad \text{and} \tag{4.42}$$

$$P_{B'} = \frac{N_{B'}}{N_P} = \frac{N_{B'}}{T}\Delta t = R_B \Delta t. \tag{4.43}$$

By inserting these values into equation (4.39), we can estimate the accidental three-fold coincidences as

$$P_{ABB'}^{(acc)} = \Delta t \left(\frac{N_{AB}}{N_A} R_{B'} + \frac{N_{AB}}{N_A} R_B \right). \tag{4.44}$$

These accidental counts will result in accidental correlations raising the value $g^{(2)}(0)$ from the predicted value of zero. The contribution from these three-fold accidental can be calculated in a straightforward manner as follows (using equations (4.28) and (4.40)),

$$g^{(2)}_{ABB'}(0)^{(acc)} = \frac{P_{ABB'}(0)^{(acc)}}{P_{AB}(0)P_{AB'}(0)} \tag{4.45}$$

$$= \Delta t \left(\frac{N_{AB}}{N_A} R_{B'} + \frac{N_{AB'}}{N_A} R_B \right) \frac{N_A}{N_{AB}} \frac{N_A}{N_{AB'}} \tag{4.46}$$

where in this last step we have used equations (4.40) and (4.41). This leads to the expression

$$g^{(2)}_{ABB'}(0)^{(acc)} = N_A \left(\frac{R_{B'}}{N_{AB'}} + \frac{R_B}{N_{AB}} \right) \tag{4.47}$$

or in terms of counts,

$$g^{(2)}_{ABB'}(0)^{(acc)} = \left(\frac{\Delta t}{T} \right) N_A \left(\frac{N_{B'}}{N_{AB'}} + \frac{N_B}{N_{AB}} \right). \tag{4.48}$$

These accidental counts are calculated for the various experimental runs. Some typical values are shown in table 4.3, and their sizes are comparable to the nonzero second-order correlation functions calculated for the quantum mechanical beams comprising of single photons.

4.3.8 Time-dependent second-order coherence

To investigate second-order coherence $g^{(2)}(\tau)$ further, we can repeat the three-detector experiment for different values of time delay τ between the two detection events. One option to add a measurement delay is by increasing the path length of one of the exit arms of the PBS, but this method is difficult to implement. So, instead of changing the *optical* settings, we introduce a delay in the *electrical* signals coming out from SPCMs by passing them through an analog delay generator[6]. According to our convention, when we install the delay box in detector B's signal, we consider it to be a positive delay, while installing it in detector B''s signal will give us a negative delay.

We sweep the delay τ from -60 ns to 60 ns in steps of 2 ns and calculate the value of $g^{(2)}$ for each delay value. Figure 4.31a shows typical $g^{(2)}(\tau)$ for bunched, coherent, and antibunched (quantum) light, whereas figure 4.31b shows the experimentally determined $g^{(2)}(\tau)$ trend obtained for our experiment. The plot in figure 4.31b illustrates a drop in the value of $g^{(2)}(\tau)$ at $\tau = 0$. This implies the antibunching of photons. In other words, a single photon is detected at either detector B or B' but not

---

[6] A delay generator is implemented by a coax delay box (Stanford Research Systems DB64).

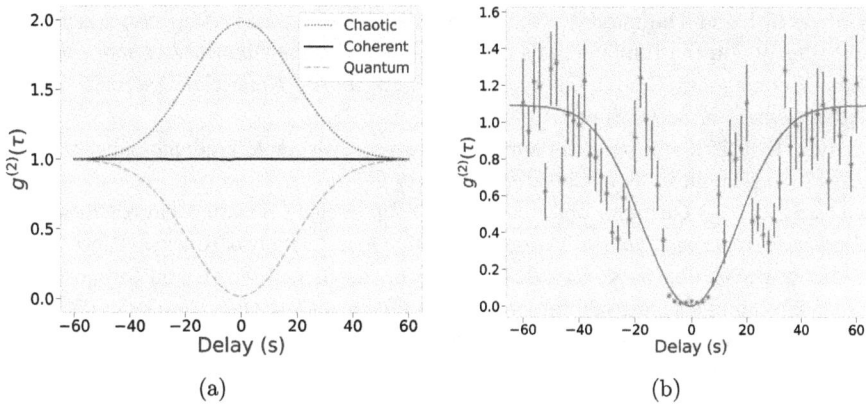

Figure 4.31. (a) Simulated $g^{(2)}(\tau)$ for chaotic, coherent, and quantum light sources. (b) Experimentally measured $g^{(2)}(\tau)$ for the single-photon source. The piecewise curve fits act as a guide to the eye. This data were acquired using a 3 ns coincidence time window.

at both simultaneously. This is *the* quantum signature of light and agrees with the simulated $g^{(2)}(\tau)$ for a quantum light source, shown in figure 4.31a.

Looking at figures 4.31a and 4.31b, one can see that the experimental $g^{(2)}(\tau)$ follows a similar trend as the simulated graphs. Our points are straightened out in the center near $\tau = 0$ because this is where our coincidence window of $\approx$3 ns lies. The two pulses that arrive within the coincidence window of our CCU are counted as coincident. So, within this window, the delay generator does not create a significant difference in the number of coincidence detections, and hence the value of $g^{(2)}(\tau)$ remains constant and very close to zero. This experiment is a photonic analog of the classic Hanbury Brown–Twiss experiment [45] and confirms the grainy nature of light.

In experiments Q1 and Q2, we have created single photons by SPDC and checked their quantum nature by quantitatively investigating $g^{(2)}(\tau)$. The results unambiguously indicate nonclassical signatures in the stream of heralded single photons. In the next chapter, we will explore experimental methods to determine the polarization states of the downconverted single photons.

Reference

[1] Beck M and Galvez E J 2007 Quantum optics in the undergraduate teaching laboratory *Conf. on Coherence and Quantum Optics* (Washington, DC: Optical Society of America) pCSuA4

[2] Galvez E J and Beck M 2007 Quantum optics experiments with single photons for undergraduate laboratories *Education and Training in Optics* (Washington, DC: Optical Society of America) pp 1–8

[3] Beck M 2012 *Quantum Mechanics: Theory and Experiment* (Oxford: Oxford University Press)

[4] Beck M and Dederick E 2013 Quantum optics laboratories for undergraduates *Education and Training in Optics and Photonics* (Washington, DC: Optical Society of America) p EWB3

[5] Galvez E J and Hamilton N Y Undergraduate laboratories using correlated photons: experiments on the fundamentals of quantum mechanics *Innovative Laboratory Design* Eds: Cunningham S and George Y S, (New York, American Association for the Advancement of Science) pp 113–8

[6] Galvez E J 2010 Correlated-photon experiments for undergraduate labs *Unpublished Handbook* (Hamilton, NY: Colgate University)

[7] Galvez E J 2019 Quantum optics laboratories for teaching quantum physics *Education and Training in Optics and Photonics* (Washington, DC: Optical Society of America) Paper 11 143_123

[8] Lukishova S G, Jha A K, Savidis N, White S, Bissell L, Elgin L and Stroud C R 2007 Quantum optics teaching laboratory *Conf. on Coherence and Quantum Optics* (Washington, DC: Optical Society of America) p JWC8

[9] Lukishova S G, Stroud C R, Bissell L, Zimmerman B and Knox W H 2008 Teaching experiments on photon quantum mechanics *Frontiers in Optics 2008/Laser Science XXIV/ Plasmonics and Meta-materials/Optical Fabrication and Testing* (Washington, DC: Optical Society of America) SThD3

[10] Lukishova S G 2017 Quantum optics and nano-optics teaching laboratory for the undergraduate curriculum: teaching quantum mechanics and nano-physics with photon counting instrumentation *Education and Training in Optics and Photonics* (Washington, DC: Optical Society of America) p 104522I

[11] Branning D, Khanal S, Shin Y H, Clary B and Beck M 2011 Note: scalable multiphoton coincidence-counting electronics *Rev. Sci. Instrum.* **82** 016102

[12] Gea-Banacloche J 2004 Optical realizations of quantum teleportation *Progress in Optics* (Amsterdam: Elsevier) vol **46** 311–54

[13] Kok P, Munro W J, Nemoto K, Ralph T C, Dowling J P and Milburn G J 2007 Linear optical quantum computing with photonic qubits *Rev. Mod. Phys.* **79** 135

[14] Scarani V, Bechmann-Pasquinucci H, Cerf N J, Dušek M, Lütkenhaus N and Peev M 2009 The security of practical quantum key distribution *Rev. Mod. Phys.* **81** 1301

[15] Aspect A, Dalibard J and Roger G 1982 Experimental test of Bell's inequalities using time-varying analyzers *Phys. Rev. Lett.* **49** 1804

[16] Gaertner S, Weinfurter H and Kurtsiefer C 2005 Fast and compact multichannel photon coincidence unit for quantum information processing *Rev. Sci. Instrum.* **76** 123108

[17] Felekyan S, Kühnemuth R, Kudryavtsev V, Sandhagen C, Becker W and Seidel C A M 2005 Full correlation from picoseconds to seconds by time-resolved and time-correlated single photon detection *Rev. Sci. Instrum.* **76** 083104

[18] Wahl M, Rahn H-J, Gregor I, Erdmann R and Enderlein J 2007 Dead-time optimized time-correlated photon counting instrument with synchronized, independent timing channels *Rev. Sci. Instrum.* **78** 033106

[19] Acremann Y, Chembrolu V, Strachan J P, Tyliszczak T and Stöhr J 2007 Software defined photon counting system for time resolved x-ray experiments *Rev. Sci. Instrum.* **78** 014702

[20] Dehlinger D and Mitchell M W 2002 Entangled photon apparatus for the undergraduate laboratory *Am. J. Phys.* **70** 898–902

[21] Branning D, Bhandari S and Beck M 2009 Low-cost coincidence-counting electronics for undergraduate quantum optics *Am. J. Phys.* **77** 667–70

[22] Masters M F, Heral T and Tummala K 2015 Low-cost coincidence counting apparatus for single photon optics investigations *2015 Conf. on Laboratory Instruction Beyond the First Year of College* (Salem, OR: Advanced Laboratory Physics Association) pp 56–9

[23] Branning D and Beck M 2012 An FPGA-based module for multiphoton coincidence counting *Advanced Photon Counting Techniques VI* **vol 8375** (Bellingham, WA: International Society for Optics and Photonics) p 83750F

[24] Park B K, Kim Y-S, Kwon O, Han S-W and Moon S 2015 High-performance reconfigurable coincidence counting unit based on a field programmable gate array *Appl. Opt.* **54** 4727–31

[25] Thorn J J, Neel M S, Donato V W, Bergreen G S, Davies R E and Beck M 2004 Observing the quantum behavior of light in an undergraduate laboratory *Am. J. Phys.* **72** 1210–9

[26] Kwiat P G, Waks E, White A G, Appelbaum I and Eberhard P H 1999 Ultrabright source of polarization-entangled photons *Phys. Rev.* A **60** R773

[27] Migdall A 1997 Polarization directions of noncollinear phase-matched optical parametric downconversion output *J. Opt. Soc. Am.* B **14** 1093–8

[28] Grangier P, Roger G and Aspect A 1986 Experimental evidence for a photon anticorrelation effect on a beam splitter: a new light on single-photon interferences *Europhys. Lett.* **1** 173

[29] Gogo A, Snyder W D and Beck M 2005 Comparing quantum and classical correlations in a quantum eraser *Phys. Rev.* A **71** 052103

[30] Ashby J M, Schwarz P D and Schlosshauer M 2016 Delayed-choice quantum eraser for the undergraduate laboratory *Am. J. Phys.* **84** 95–105

[31] Altepeter J B, Jeffrey E R and Kwiat P G 2005 Photonic state tomography *Adv. Atom. Mol. Opt. Phy.* **52** 105–59

[32] Brody J and Selton C 2018 Quantum entanglement with Freedman's inequality *Am. J. Phys.* **86** 412–6

[33] Dehlinger D and Mitchell M W 2002 Entangled photons, nonlocality, and Bell inequalities in the undergraduate laboratory *Am. J. Phys.* **70** 903–10

[34] Carlson J A, Olmstead M D and Beck M 2006 Quantum mysteries tested: an experiment implementing Hardy's test of local realism *Am. J. Phys.* **74** 180–6

[35] Galvez E J, Holbrow C H, Pysher M J, Martin J W, Courtemanche N, Heilig L and Spencer J 2005 Interference with correlated photons: five quantum mechanics experiments for undergraduates *Am. J. Phys.* **73** 127–40

[36] Lamb W E Jr 1995 Anti-photon *Appl. Phys.* B **60** 77–84

[37] Roychoudhuri C E, Kracklauer A F and Creath K 2024 *The Nature of Light* (Boca Raton, FL: CRC Press)

[38] Loudon R 2010 *The Quantum Theory of Light* (Oxford: Oxford University Press)

[39] Stanley R Q 1996 Question # 45. What (if anything) does the photoelectric effect teach us? *Am. J. Phys.* **64** 839

[40] Milonni P W 1997 Answer to question # 45 ['What (if anything) does the photoelectric effect teach us?', RQ Stanley, Am. J. Phys. 64 (7), 839 (1996) *Am. J. Phys.* **65** 11–2

[41] Burnham D C and Weinberg D L 1970 Observation of simultaneity in parametric production of optical photon pairs *Phys. Rev. Lett.* **25** 84

[42] Clauser J F 1974 Experimental distinction between the quantum and classical field-theoretic predictions for the photoelectric effect *Phys. Rev.* D **9** 853

[43] Kimble H J, Dagenais M and Mandel L 1977 Photon antibunching in resonance fluorescence *Phys. Rev. Lett.* **39** 691

[44] Funk A C and Beck M 1997 Sub-Poissonian photocurrent statistics: theory and undergraduate experiment *Am. J. Phys.* **65** 492–500

[45] Brown R H, Twiss R Q *et al* 1956 Correlation between photons in two coherent beams of light *Nature* **177** 27–9

[46] Twiss R Q, Little A G and Brown R H 1957 Correlation between photons, in coherent beams of light, detected by a coincidence counting technique *Nature* **180** 324

[47] Glauber R J 1963 Photon correlations *Phys. Rev. Lett.* **10** 84

[48] Glauber R J 1963 The quantum theory of optical coherence *Phys. Rev.* **130** 2529

[49] Glauber R J 1963 Coherent and incoherent states of the radiation field *Phys. Rev.* **131** 2766

[50] Sudarshan E C G 1963 Equivalence of semiclassical and quantum mechanical descriptions of statistical light beams *Phys. Rev. Lett.* **10** 277

[51] Kelley P L and Kleiner W H 1964 Theory of electromagnetic field measurement and photoelectron counting *Phys. Rev.* **136** A316

[52] Mandel L and Wolf E 1965 Coherence properties of optical fields *Rev. Mod. Phys.* **37** 231

[53] Arecchi F T, Gatti E and Sona A 1966 Time distribution of photons from coherent and Gaussian sources *Phys. Lett.* **20** 27–9

[54] Clauser J F 1972 Experimental limitations to the validity of semiclassical radiation theories *Phys. Rev.* A **6** 49

[55] Fox M 2006 *Quantum Optics: An Introduction* (Oxford: Oxford University Press)

[56] Zou X T and Mandel L 1990 Photon-antibunching and sub-Poissonian photon statistics *Phys. Rev.* A **41** 475

[57] Greenstein G and Zajonc A 2006 *The Quantum Challenge: Modern Research on the Foundations of Quantum Mechanics* (Burlington, MA: Jones & Bartlett Learning)

[58] Blay D R, Steel M J and Helt L G 2017 Effects of filtering on the purity of heralded single photons from parametric sources *Phys. Rev.* A **96** 053842

[59] Graffitti F, Kelly-Massicotte J, Fedrizzi A and Brańczyk A M 2018 Design considerations for high-purity heralded single-photon sources *Phys. Rev.* A **98** 053811

[60] Pearson B J and Jackson D P 2010 A hands-on introduction to single photons and quantum mechanics for undergraduates *Am. J. Phys.* **78** 471–84

[61] Eckart C and Shonka F R 1938 Accidental coincidences in counter circuits *Phys. Rev.* **53** 752–6

IOP Publishing

Quantum Mechanics in the Single-Photon Laboratory (Second Edition)

Muhammad Sabieh Anwar, Faizan-e-Ilahi, Syed Bilal Hyder and Muhammad Hamza Waseem

Chapter 5

The polarization of photons

Photons are the quantum of light. Each photon is a complete entity carrying all the polarization details relevant to an electromagnetic field. We have covered polarization in some level of detail in chapter 2. In the previous chapter, we were mainly concerned with photodetection statistics and the grainy nature of light itself, indicating that photons really exist. Moving further on this theme, the experiments covered in this chapter will highlight techniques to *measure* the polarization of single photons. These experiments are listed below:

- Q3: Estimating the polarization state of single photons
- Q4: Visualizing the polarization state of single photons

In these experiments, we will perform measurements on the signal beam of down-converted photons. Unlike a steady laser beam that we can analyze repeatedly for its polarization, a photon can only be measured once. The act of measurement consumes the photon. Therefore, to determine the polarization, we need to perform measurements on thousands of photons. The detection statistics then help us to unveil the polarization properties of the photons.

For the current chapter, we are assuming that the photons are in a pure state and can be represented using a state vector or a ket. However, as discussed in chapter 3, a more general representation of an arbitrary quantum state, a density matrix, also exists. We postpone the experimental investigation of estimating the density matrix of an arbitrary polarization-encoded state until chapter 8. At that point, we will conduct an experiment (which we call quantum state tomography, or QST) to determine the complete density matrix, without putting any constraints on the state purity.

doi:10.1088/978-0-7503-6315-0ch5 © IOP Publishing Ltd 2024

5.1 Q3: Estimating the polarization state of single photons

Experiment Q3 follows the method presented in [1] and will be revisited from alternative perspectives in experiments Q4 and QST. Just as in the previous experiments, the idler photons will be used to herald the detection of signal photons. We are not interested in the polarization state of the idler photons but only the polarization state of the signal photon.

We start by reviewing a method to estimate the polarization state of an ensemble of identically prepared single photons. The quantum state of the signal photons encodes the polarization state and, as discussed in chapter 3, can be generally represented as

$$|\psi\rangle = A|H\rangle + Be^{i\phi}|V\rangle, \tag{5.1}$$

where $A^2 + B^2 = 1$, and A, B, and φ are real numbers. So our task boils down to determining the three parameters A, B, and ϕ. To accomplish this, we perform a large number of polarization measurements on ensembles of photons that are identically prepared in the state $|\psi\rangle$. We use three bases $\{|H\rangle, |V\rangle\}$, $\{|D\rangle, |A\rangle\}$ and $\{|L\rangle, |R\rangle\}$ to perform these measurements so that the probabilities $P_{|H\rangle}$, $P_{|D\rangle}$, and $P_{|L\rangle}$ can be determined. We employ these probabilities to compute the real-valued parameters A, B, and ϕ. The notation $P_{|H\rangle}$ is short for $P(|H\rangle||\psi\rangle)$, i.e., the probability to project $|\psi\rangle$ onto the state $|H\rangle$, and so on for the other configurations.

If we take a beam of photons identically generated in the state $|\psi\rangle$ and perform a large number of polarization measurements in the canonical basis $\{|H\rangle, |V\rangle\}$, according to Born's rule [2], we can determine the probability of detecting a horizontally polarized photon as

$$P_{|H\rangle} = |\langle H|\psi\rangle|^2 = A^2. \tag{5.2}$$

From this probability and the normalization constraint on the state vector, we can actually obtain both A and B as

$$A = \sqrt{P_{|H\rangle}} \quad \text{and} \tag{5.3}$$

$$B = \sqrt{1 - A^2}. \tag{5.4}$$

Once we determine A and B, we would like to determine ϕ which requires a similar set of measurements but in the $\{|D\rangle, |A\rangle\}$ and $\{|L\rangle, |R\rangle\}$ bases. It is important that the state-generation setup does not change during all these measurements because we do not want to modify the state $|\psi\rangle$ until it has been probed by all the measurements in the three bases.

We change the measurement basis to $\{|D\rangle, |A\rangle\}$ and perform a large number of measurements. Once again, we can compute the probability of detecting a diagonally polarized photon using Born's rule as

$$P_{|D\rangle} = |\langle D|\psi\rangle|^2$$
$$= |\langle D|(A|H\rangle + Be^{i\phi}|V\rangle)|^2 \tag{5.5}$$
$$= \frac{1 + 2AB\cos\phi}{2}.$$

Since we have already determined both A and B, we can invert equation (5.5) to obtain $\cos\phi$ from $P(|D\rangle||\psi\rangle)$. However, this does not uniquely determine ϕ because the inverse of cosine is not unique. For example, $\cos(\frac{\pi}{3}) = \cos(\frac{-\pi}{3})$. Hence, measurements in a third basis must be performed on photons generated in the same state. Therefore, we modify the scheme to perform measurements in the $\{|L\rangle, |R\rangle\}$ basis. Now, the probability of detecting a left circularly polarized photon is given by

$$P_{|L\rangle} = |\langle L|\psi\rangle|^2$$
$$= |\langle L|(A|H\rangle + Be^{i\phi}|V\rangle)|^2 \tag{5.6}$$
$$= \frac{1 + 2AB\sin\phi}{2}.$$

We can manipulate this equation to obtain $\sin\phi$. Finally, knowledge of both $\cos\phi$ and $\sin\phi$ lets us uniquely determine ϕ. In fact, one can obtain the following useful expression for ϕ by manipulating equations (5.5) and (5.6):

$$\phi = \tan^{-1}\left(\frac{P_{|L\rangle} - \frac{1}{2}}{P_{|D\rangle} - \frac{1}{2}}\right). \tag{5.7}$$

5.1.1 Generating polarization states

As in experiments Q1 and Q2, detecting an idler photon is essential to prepare a single-photon state in the signal beam. An arbitrary polarization of this single-photon state can be generated simply by adding a half-wave plate (HWP) or a quarter-wave plate (QWP) in the path of the signal beam right after the downconversion crystal. So, if the downconversion produces horizontally polarized photons, this HWP can generate any linear polarization. We borrow the transformation matrix $\hat{O}_{\mathrm{HWP}}$ for this operation from table 3.2 and compute the generated state as

$$|\psi\rangle = \hat{O}_{\mathrm{HWP}}|H\rangle$$
$$= \begin{pmatrix} \cos 2h & \sin 2h \\ \sin 2h & -\cos 2h \end{pmatrix}\begin{pmatrix} 1 \\ 0 \end{pmatrix} \tag{5.8}$$
$$= \begin{pmatrix} \cos 2h \\ \sin 2h \end{pmatrix},$$

where we specify h as the angle of the HWP optic axis with the horizontal. From equation (5.8), it can be shown that with HWP orientations $h = 0°$, $45°$, $22.5°$, $-22.5°$, states $|H\rangle$, $|V\rangle$, $|D\rangle$, and $|A\rangle$ can be generated when input with the $|H\rangle$ state. Hence, we can label this HWP as the 'state-generation' HWP. On the other hand, to generate circular or elliptical polarization, a state-generation QWP is used instead of the state-generation HWP. Once again, we utilize the transformation matrix $\hat{O}_{QWP}$ from table 3.2 and compute the generated state as

$$
\begin{aligned}
|\psi\rangle &= \hat{O}_{QWP}|H\rangle \\
&= \begin{pmatrix} \cos^2 q + i\sin^2 q & (1-i)\sin q \cos q \\ (1-i)\sin q \cos q & \sin^2 q + i\cos^2 q \end{pmatrix}\begin{pmatrix} 1 \\ 0 \end{pmatrix} \\
&= c\begin{pmatrix} \cos^2 q + i\sin^2 q \\ (1-i)\sin q \cos q \end{pmatrix},
\end{aligned}
\tag{5.9}
$$

where q now specifies the orientation of the QWP fast axis with respect to the horizontal. From equation (5.9), it can be shown that states $|L\rangle$ and $|R\rangle$ can be generated with $q = 45°$ and $-45°$ respectively.

5.1.2 Measuring polarization states

Once a polarization state is generated, it needs to be measured. The measurement always refers to some basis. Intuitively, the idea of a measurement basis can be understood as follows. If we want to measure in the $\{|H\rangle, |V\rangle\}$ basis, then the measuring apparatus should be able to distinguish between $|H\rangle$ and $|V\rangle$ polarized light. Measurements in this basis can be readily performed using the PBS, which transmits vertically polarized photons and deflects horizontally polarized photons through $90°$ to a different trajectory. The probability that an individual photon will transmit through the PBS and thus fall at detector B is equal to the probability of registering an AB coincidence event. This probability can be estimated from the measured coincidence counts as follows:

$$
P_{AB} = \frac{N_{AB}}{N_{AB} + N_{AB'}}.
\tag{5.10}
$$

Similarly, the PBS reflects the horizontally polarized photons, which then travel toward detector B', and hence the probability that an individual photon is reflected off the PBS is given by

$$
P_{AB'} = \frac{N_{AB'}}{N_{AB} + N_{AB'}}.
\tag{5.11}
$$

This configuration is presented in 5.1.

Figure 5.1 helps us understand the measurement process better. Placing a PBS in the path of an incoming photon in state $|\psi\rangle$ implements the measurement operator $\hat{M}_{HV} = |H\rangle\langle H| - |V\rangle\langle V|$. The successive probabilities $P_{|H\rangle}$ and $P_{|V\rangle}$ are inferred from the counts N_H and N_V received from the detectors D_H and D_V. The predictions are given by equation (5.2),

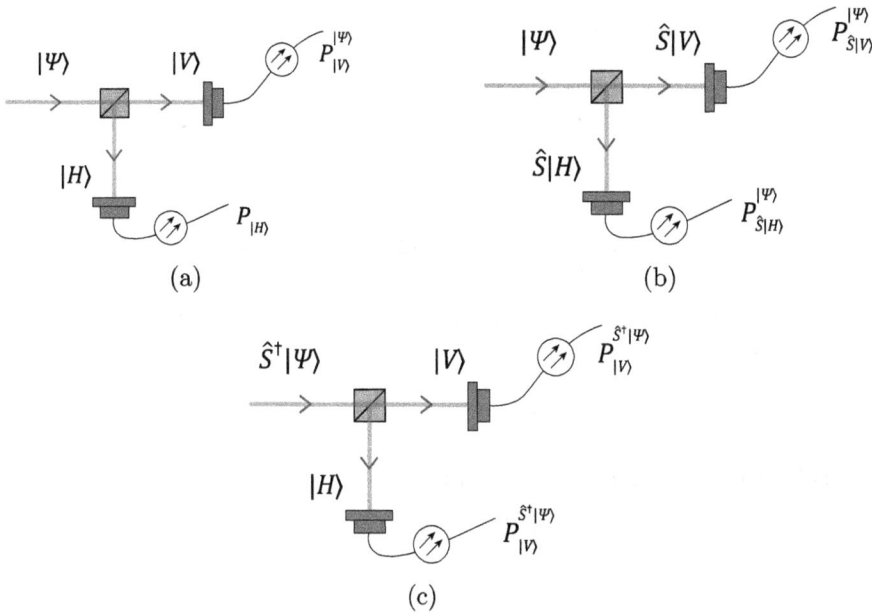

(a)　　　　　　　　　　　　　　(b)

(c)

Figure 5.1. (a) Without any basis rotation, the PBS differentiates between $|H\rangle$ and $|V\rangle$ polarizer photons by transmitting one and reflecting the other. (b) Within the rotated basis, the PBS is supposed to differentiate between two new orthogonal states that have evolved from $|H\rangle$ and $|V\rangle$ due to a basis change operator $\hat{S}$. (c) With the $\hat{S}^{\dagger}$ operator, we rotate the state itself to match the rotated measurement basis. For the PBS, separating this state into two orthogonal components will be equivalent to distinguishing between $\hat{S}|H\rangle$ and $\hat{S}|V\rangle$.

$$P_{|H\rangle} = |\langle H|\psi\rangle|^2 \quad \text{and}$$
$$P_{|V\rangle} = |\langle V|\psi\rangle|^2$$

which can also be written as:

$$P_{|H\rangle} = \langle\psi|H\rangle\langle H|\psi\rangle = \langle|H\rangle\langle H|\rangle_{|\psi\rangle} \qquad \text{and} \tag{5.12}$$

$$P_{|V\rangle} = \langle\psi|V\rangle\langle V|\psi\rangle = \langle|V\rangle\langle V|\rangle_{|\psi\rangle}, \tag{5.13}$$

where we borrow notation from quantum physics which specifies $\langle|H\rangle\langle H|\rangle_{|\psi\rangle}$ as the expectation value of the operator $|H\rangle\langle H|$ when the quantum state is $|\psi\rangle$ and likewise for $\langle|V\rangle\langle V|\rangle_{|\psi\rangle}$. For more clarity, we write the probabilities in equations (5.12) and (5.13) as

$$P_{|H\rangle}^{|\psi\rangle} = \langle|H\rangle\langle H|\rangle_{|\psi\rangle} \quad \text{and} \tag{5.14}$$

$$P_{|V\rangle}^{|\psi\rangle} = \langle|V\rangle\langle V|\rangle_{|\psi\rangle}, \tag{5.15}$$

where the measurand state $|\psi\rangle$ is identified in a superscript. The expectation value of the measurement operator $\hat{M}_{HV}$ provides us the net difference in probabilities expressed through the following operation

$$P_{|H\rangle}^{|\psi\rangle} - P_{|V\rangle}^{|\psi\rangle} = \langle\psi|(|H\rangle\langle H| - |V\rangle\langle V|)|\psi\rangle \tag{5.16}$$

$$= \langle|H\rangle\langle H| - |V\rangle\langle V|\rangle_{|\psi\rangle} \tag{5.17}$$

$$= \langle\hat{M}_{HV}\rangle_{|\psi\rangle}. \tag{5.18}$$

Now suppose we wish to make a measurement in the rotated basis $\{|\phi\rangle, |\phi^\perp\rangle\}$ defined through the basis transformation matrix, $\hat{S}$, with the rotated states[1] given by $|\phi\rangle = \hat{S}|H\rangle$ and $|\phi^\perp\rangle = \hat{S}|V\rangle$. This task is illustrated in figure 5.1b. For example, consider measurement in the $\{|\phi\rangle = |D\rangle, |\phi^\perp\rangle = |A\rangle\}$ basis, which is produced through the transformation matrix

$$\hat{S} = \frac{1}{\sqrt{2}}\begin{pmatrix} 1 & 1 \\ 1 & -1 \end{pmatrix}. \tag{5.19}$$

The target is to achieve the measurement

$$\hat{M}_{\hat{S}|H\rangle, \hat{S}|V\rangle} = \hat{S}|H\rangle\langle H|\hat{S}^\dagger - \hat{S}|V\rangle\langle V|\hat{S}^\dagger \tag{5.20}$$

$$= \hat{S}(|H\rangle\langle H| - |V\rangle\langle V|)\hat{S}^\dagger, \tag{5.21}$$

and thus obtain the net difference in probabilities

$$P_{\hat{S}|H\rangle}^{|\psi\rangle} - P_{\hat{S}|V\rangle}^{|\psi\rangle} = \langle\psi|\hat{M}_{\hat{S}|H\rangle, \hat{S}|V\rangle}|\psi\rangle \tag{5.22}$$

$$= \langle\psi|\hat{S}(|H\rangle\langle H| - |V\rangle\langle V|)\hat{S}^\dagger|\psi\rangle \tag{5.23}$$

$$= ((\langle\psi|\hat{S})(|H\rangle\langle H| - |V\rangle\langle V|)(\hat{S}^\dagger|\psi\rangle)) \tag{5.24}$$

$$= P_{|H\rangle}^{\hat{S}^\dagger|\psi\rangle} - P_{|V\rangle}^{\hat{S}^\dagger|\psi\rangle}. \tag{5.25}$$

The last expression shows that the probabilities in the rotated basis $\{|\psi\rangle, |\psi^\perp\rangle\}$ can be computed from the probabilities that are recovered from the *rotated state* $\hat{S}^\dagger|\psi\rangle$, measured in the *unrotated basis* $\{|H\rangle, |V\rangle\}$. Therefore, instead of changing the orientation of the measurement, one may *bring the state* to a rotated configuration instead! The state rotation strategy is highlighted in figure 5.1c, which achieves exactly the same effect as the requirement spelled out in figure 5.1b.

In short, for measurement in the $\{|D\rangle, |A\rangle\}$ basis, guided by equation (5.21), we must implement the measurement operator $\hat{M}_{DA}$ which is given by,

$$\hat{S}(|H\rangle\langle H| - |V\rangle\langle V|)\hat{S}^\dagger = |D\rangle\langle D| - |A\rangle\langle A|. \tag{5.26}$$

This requires the transformation,

$$\hat{S} = \frac{1}{\sqrt{2}}\begin{pmatrix} 1 & 1 \\ 1 & -1 \end{pmatrix} \quad \text{and} \tag{5.27}$$

---

[1] In quantum mechanics, the matrix $\hat{S}$ is called a similarity transform.

$$\hat{S}^\dagger = \frac{1}{\sqrt{2}}\begin{pmatrix} 1 & 1 \\ 1 & -1 \end{pmatrix}. \tag{5.28}$$

How do we practically implement the $\hat{S}^\dagger$ operator? Table 2.2 shows that the transformation equation (5.28) can be achieved with a HWP set at $\theta = h = 22.5°$. Similarly, for measurement in the $\{|L\rangle, |R\rangle\}$ basis, we require,

$$\hat{S}(|H\rangle\langle H| - |V\rangle\langle V|)\hat{S}^\dagger = |L\rangle\langle L| - |R\rangle\langle R|, \tag{5.29}$$

warranting the transformation matrices,

$$\hat{S} = \frac{1}{\sqrt{2}}\begin{pmatrix} 1 & i \\ i & 1 \end{pmatrix} \quad \text{and} \tag{5.30}$$

$$\hat{S}^\dagger = \frac{1}{\sqrt{2}}\begin{pmatrix} 1 & -i \\ -i & 1 \end{pmatrix}. \tag{5.31}$$

Table 2.2 shows that equation (5.31) is implemented by a QWP with the orientation $\theta = q = 45°$. Therefore, one way to perform measurements along the various bases is by placing, one by one, a HWP or a QWP before the PBS. The desired setting is shown in table 5.1, and is also illustrated in figure 5.2a.

However, plugging in and out a wave plate is cumbersome and may destroy the precious alignment. It turns out that it is also possible to keep a QWP together with

Table 5.1. Commonly used measurement basis for polarization-encoded quantum state estimation. In this scheme, one places either a HWP or a QWP at a particular angle to change the measurement basis. Note that the HWP is optional for the $\{|H\rangle, |V\rangle\}$ measurement basis.

Measurement basis	QWP	HWP	q	h
$\{|H\rangle, |V\rangle\}$	X	✓	N.A.	0°
$\{|D\rangle, |A\rangle\}$	X	✓	N.A.	22.5°
$\{|L\rangle, |R\rangle\}$	✓	X	45°	N.A.

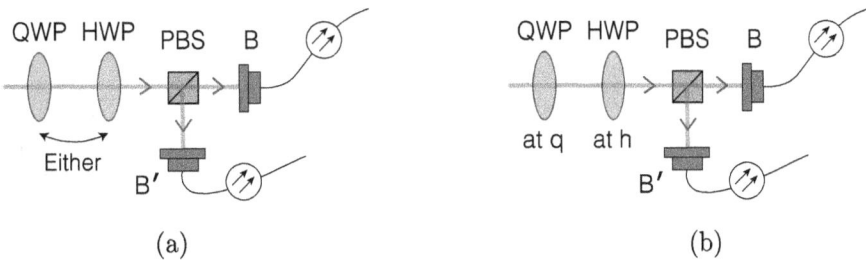

(a)　　　　　　　　　　　　　　(b)

Figure 5.2. (a) The arrangement to change the measurement basis using one wave plate at a time. (b) Arrangement to change the measurement using the two wave plates simultaneously kept in place, a QWP followed by a HWP.

Table 5.2. Alternative scheme for basis change. In this scheme, a QWP followed by a HWP is inserted in the beam path. Using particular angles generates the desired measurement basis.

Measurement basis	QWP	HWP	q	h
$\{|H\rangle, |V\rangle\}$	✓	✓	$0°$	$0°$
$\{|D\rangle, |A\rangle\}$	✓	✓	$45°$	$22.5°$
$\{|L\rangle, |R\rangle\}$	✓	✓	$45°$	$0°$

and *followed by* a HWP, and adjust their angles q and h for measuring along the various bases. This scenario is shown in figure 5.2b and the required settings of q and h are presented in table 5.2.

One can check that these arrangements of the QWP and the HWP specified in table 5.2 do indeed work. Insert the values of q and h into $\hat{J}_{QWP}(q)$ and $\hat{J}_{HWP}(h)$ provided in table 2.2 and compute the matrix $(\hat{J}_{HWP}(h)\hat{J}_{QWP}(q))$. Set it equal to $\hat{S}^\dagger$ and verify that $\hat{S}(|H\rangle\langle H| - |V\rangle\langle V|)\hat{S}^\dagger$ yields the correct measurement operator, $\hat{M}_{HV}$, $\hat{M}_{DA}$, or $\hat{M}_{LR}$. Let us do a quick check. Suppose that we are required to measure in the $\{|D\rangle, |A\rangle\}$ basis. We can look up the angles in the second row in table 5.2, and calculate

$$\hat{S}^\dagger = \hat{J}_{HWP}(h = 22.5°)\hat{J}_{QWP}(q = 45°) \tag{5.32}$$

$$= \frac{1}{\sqrt{2}}\begin{pmatrix} 1 & 1 \\ i & -i \end{pmatrix}, \tag{5.33}$$

and from hereon we can immediately show that

$$\hat{S}(|H\rangle\langle H| - |V\rangle\langle V|)\hat{S}^\dagger = \frac{1}{\sqrt{2}}\begin{pmatrix} 1 & -i \\ 1 & i \end{pmatrix}\begin{pmatrix} 1 & 0 \\ 0 & -1 \end{pmatrix}\frac{1}{\sqrt{2}}\begin{pmatrix} 1 & 1 \\ i & -i \end{pmatrix} \tag{5.34}$$

$$= \begin{pmatrix} 0 & 1 \\ 1 & 0 \end{pmatrix} \tag{5.35}$$

which is equal to $|D\rangle\langle D| - |A\rangle\langle A|$, where all our matrices are written in the $\{|H\rangle, |V\rangle\}$ representation. The angles q and h are hence correct.

We can approach this problem from yet one more standpoint. An arbitrary state $|\psi\rangle = A|H\rangle + Be^{i\phi}|V\rangle$ gets transformed by the matrix in equation (5.33) to,

$$\hat{S}^\dagger|\psi\rangle = \frac{1}{\sqrt{2}}\begin{pmatrix} A + Be^{i\phi} \\ i(A - Be^{i\phi}) \end{pmatrix} \tag{5.36}$$

$$= \frac{1}{\sqrt{2}}(A + Be^{i\phi})|H\rangle + \frac{i}{\sqrt{2}}(A - Be^{i\phi})|V\rangle \tag{5.37}$$

and for this state the probability of detection in $|H\rangle$ appearing on one arm of the PBS output is,

$$\left| \langle H | \hat{S}^{\dagger} | \psi \rangle |^2 = \left| \frac{1}{\sqrt{2}} (A + Be^{i\phi}) \right|^2 \right. \tag{5.38}$$

$$= \frac{1 + 2AB\cos(\phi)}{2}, \tag{5.39}$$

perfectly identical to the $P_{|D\rangle}$ derived in equation (5.5).

5.1.3 The experiment

The layout of this experiment is similar to that of experiment Q2 with a few additional wave plates in the signal beam. All the wave plates between the downconversion crystals and the detectors are mounted in motorized rotation stages so that we can use the computer to accurately orient the plates. A schematic is drawn in figure 5.3 and a photograph is shown in figure 5.4.

The downconversion crystal and the detectors are aligned following the procedures outlined in experiments Q1 and Q2. The laser is vertically polarized, and the HWP in the pump beam is also oriented at 0°, leaving the pump polarization unchanged and resulting in downconverted $|HH\rangle$ photons. Note that the HWP could also be oriented at an angle of 45° with respect to the vertical, resulting in $|VV\rangle$ downconverted photons, but this is a matter of choice.

Different polarization states are generated in the signal beam using the state-generation HWP or the QWP. The HWP creates linearly polarized states with arbitrary orientation, whereas the QWP creates elliptically/circularly polarized states. Polarization measurements of identically prepared photons are performed in the three different bases, namely $\{|H\rangle, |V\rangle\}$, $\{|D\rangle, |A\rangle\}$ and $\{|L\rangle, |R\rangle\}$ by using the single wave-plate scheme summarized by table 5.1. Detection probabilities are determined using detection statistics and equations (5.2), (5.5), and (5.6). Alternatively, we could use the two double wave-plate scheme in table 5.2, but the results from this scheme are not shown here and will rather be retouched in chapter 8.

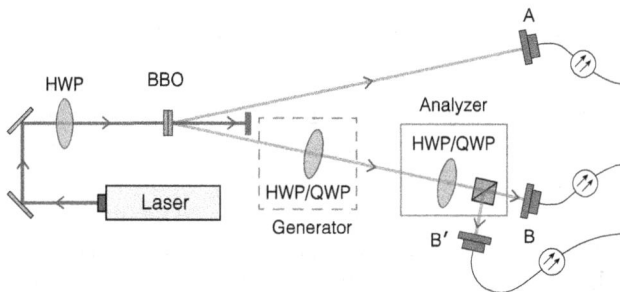

Figure 5.3. Schematic diagram of the quantum state measurement experiment. The first dotted box represents the polarization state generator. It comprises a wave plate (QWP or HWP) and changes the polarization of downconverted $|H\rangle$ photons. The second dotted box represents a polarization state analyzer and includes a QWP and/or a HWP, and a polarizing beam splitter (PBS), followed by the three detectors A, B, and B'.

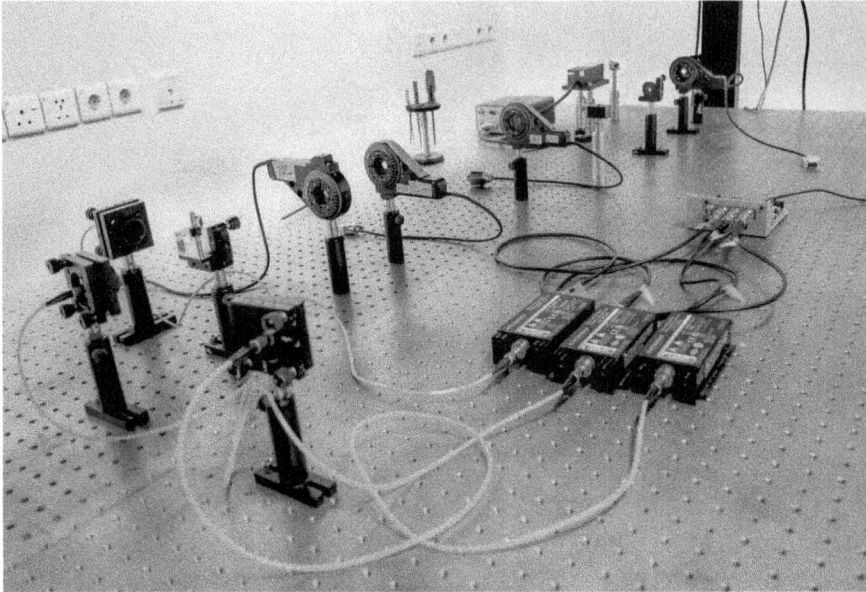

Figure 5.4. Photograph of the polarization measurement experiment, which is experiment Q3.

The data acquisition time for measurement in each basis is set at two minutes. Parameters of the polarization-encoded quantum state, A, B, and φ, are determined using equations (5.3), (5.4), (5.5), and (5.6). The experimental results show reasonable agreement with theoretical predictions as summarized in table 5.3. The deviations of measurement from predictions are due to our inability to perfectly prepare the desired states, as well as errors in the measurement basis. These imperfect conditions mainly stem from errors in orienting the wave plates in the pump and signal beams. Furthermore, the theoretical predictions rely on the assumption that we are generating pure states, which is not necessarily the case. This experiment will be revisited in a new light when we discuss experiments Q4 and QST, and where the concept of state purity will be explored further.

5.2 Q4: 'Visualizing' the polarization state of single photons

In chapter 2, we used Jones calculus to mathematically express the polarization of light, as well as compute the effects of optical elements, such as polarizers and wave plates on the polarization [3, 4]. For single-photon systems, we used a mathematical formalism analogous to Jones calculus. Experiment Q3 exploited this formation too. We know that Jones calculus is based on the amplitude picture of polarized light, i.e., the electric field. However, the photodetectors monitor light intensity, not amplitude. In chapter 2, we briefly discussed the Stokes polarization parameters, which enable us to characterize the polarization state of light through the four *intensity* parameters [4], S_0, S_1, S_2, and S_3. These parameters are often used to characterize the polarization of an optical beam due to the fact that the polarization ellipse [4] is not *directly* measurable [5]. Still there exists a work around based on the observation that the

Table 5.3. Predicted and measured results of polarization state measurement of single photons. All angles, represented by φ, are quoted in radians. The values in parentheses represent uncertainties in the last digits. The phases for the $|H\rangle$ and $|V\rangle$ states are undefined.

Input	Prediction	Measurement
$|H\rangle$	$A = 1.000$ $B = 0.000$	$A = 0.993(2),$ $B = 0.120(2)$ $\phi = 1.652(7)$
$|V\rangle$	$A = 0.000$ $B = 1.000$	$A = 0.254(2),$ $B = 0.967(3)$ $\phi = -1.156(1)$
$|D\rangle$	$A = 0.707$ $B = 0.707$ $\phi = 0.000$	$A = 0.748(2)$ $B = 0.663(2)$ $\phi = 0.201(3)$
$|A\rangle$	$A = 0.707$ $B = 0.707$ $\phi = 3.142$	$A = 0.698(1)$ $B = 0.716(1)$ $\phi = 3.055(2)$
$|L\rangle$	$A = 0.707$ $B = 0.707$ $\phi = 1.571$	$A = 0.732(1)$ $B = 0.681(1)$ $\phi = 1.490(4)$
$|R\rangle$	$A = 0.707$ $B = 0.707$ $\phi = -1.571$	$A = 0.663(2)$ $B = 0.749(2)$ $\phi = -1.530(5)$

key ellipse parameters are expressible in terms of Stokes parameters, and vice versa [4]. You may be motivated to see equation (2.19), which precisely spells out the connection between the polarization ellipse and the Stokes parameters.

In this section, we demonstrate a graphical technique to determine and visualize the polarization of single photons and compare the results with those obtained in experiment Q3. This experiment was originally published in reference [6]. It is a quantum optical version of the experiment C2 performed in chapter 2 and is motivated by the experiment discussed in reference [7]. The graphical method is based on well-established techniques used in telecommunications and electrical engineering for determining polarization profile of antennas. So, while experiment Q3 estimates the probability amplitudes and polarization parameters, Q4 helps *visualize* these properties. We will now briefly review the underlying technique.

5.2.1 Antenna polarimetry and the polarization pattern method

There exists a time-tested empirical technique to determine the polarization of antennas, known as the 'polarization pattern method' [8]. This method involves the use of a linearly polarized antenna to receive electromagnetic radiation that is being transmitted from another antenna whose polarization profile we are interested in. The receiving antenna is positioned at a normal to the propagation direction of the radiation. The receiving antenna is rotated in a number of steps, and the received signal is recorded against the various orientations of the receiving antenna. A peanut-shaped pattern is obtained when one makes a polar plot of the square root of the signal versus the receiving antenna's rotation. This is exemplified in figure 5.5. The pattern circumscribes the polarization ellipse. This peculiar shape is mathematically classified as a hippopede, also known as the 'horse-fetter' [9, 10].

From the figure, it is evident that both the hippopede and the ellipse have the same maximum and minimum radial points in the polarization ellipse. Moreover, both closed figures have the same axial ratio (AR) as well as the orientation angle (ψ). The AR is the ratio of semimajor and semiminor axes. Therefore, by making the corresponding hippopedal plot, one can determine AR and ψ, which are essential parameters required to characterize an arbitrary polarization.

However, the described method does not fetch us information about the handedness or the helicity of the polarized radiation. To obtain the helicity information, the receiving antenna needs to be replaced, one by one, by a left-hand circularly polarized antenna and then a right-hand circularly polarized antenna. The antennas should be of equal gain, and the received signal is recorded for both receiver settings. Comparing the two measured signals helps one determine the helicity of the input polarization. For instance, if the signal recorded by the right-handed setting is larger

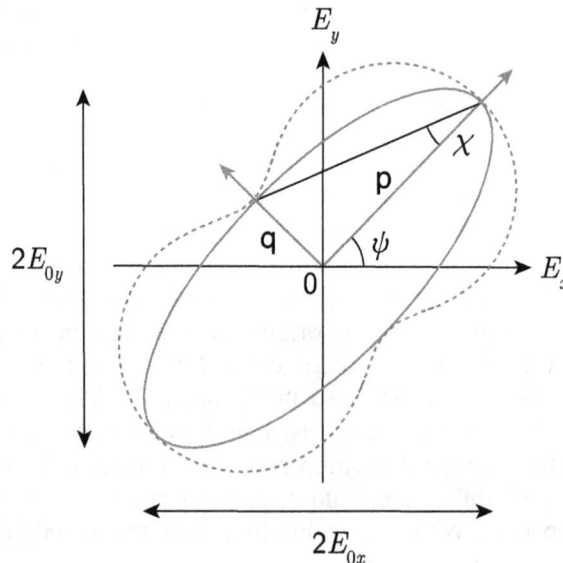

Figure 5.5. Relation of the polarization ellipse and the hippopedal pattern for an arbitrary polarization state. Angular parameters of the ellipse include the orientation angle ψ and the ellipticity angle χ. The ellipticity angle is related to the ratio of the semimajor and semiminor axes, denoted as p and q, respectively. The generation of the hippopede is described in the main text. Adapted with permission from [6]. © The Optical Society.

than the left-handed setting, then it implies that the transmitting antenna has a right-handed polarization [11].

5.2.2 Polarization pattern of single photons

For defining the relevant parameters of polarization, it is useful to carefully observe the polarization ellipse shown in figure 5.5. We can express an arbitrary ellipse using two angular parameters [4]. These parameters are the orientation angle ψ, which can be described as the tilt of the ellipse, and the ellipticity angle χ, which can be determined from the AR of the ellipse. Another angular parameter is the auxiliary angle, which is defined as $\gamma = \tan^{-1}(E_{0y}/E_{0x})$, where E_{0x} and E_{0y}, respectively, denote the maximal amplitudes of the x and y components of the polarization. This parameter can be used in an alternative angular description of the ellipse.

The polarization pattern method for antennas can be readily adapted to an optical system. In the antenna system, a transmitting antenna was responsible for the polarized radiation. For the optical version of the experiment, we can use a scheme similar to that of experiment Q3 for generating an optical beam of arbitrary polarization. Furthermore, instead of using an antenna as an analyzer, we utilize an optical polarization analyzer. Then, it is also straightforward to develop a correspondence between a linearly polarized antenna and an optical analyzer consisting of a linear polarizer and a photodetector. Finally, the optical replacement of a circularly polarized receiving antenna can be a QWP with its axis oriented at $\pm 45°$, followed by a horizontally oriented linear polarizer and a photodetector.

Let us look at this analogy mathematically. Take an optical beam of arbitrarily polarized light, which is prepared by a state generator (figure 5.6). If we consider the beam to consist of single photons, as in experiment Q3, we can say that the polarization state encodes the quantum state of single photons, represented by

$$|\psi\rangle = \cos\gamma|H\rangle + \sin\gamma\, e^{i\phi}|V\rangle. \qquad (5.40)$$

Here, the parameters γ and φ suffice to give a complete characterization of the quantum state. Equation (5.40) is merely another way of writing equation (3.19), and the correspondence is clear, $A = \cos\gamma$, $B = \sin\gamma$, and $A^2 + B^2 = 1$.

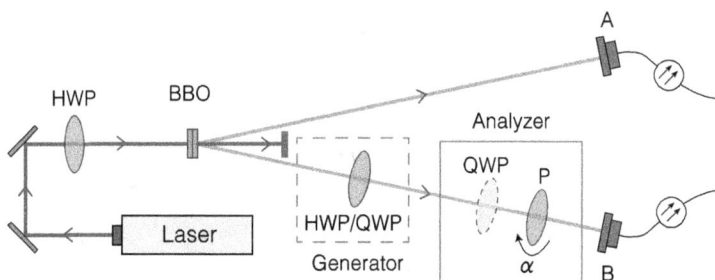

Figure 5.6. Schematic diagram of the polarization visualization experiment. The dashed box represents the polarization state generator and includes a HWP and a QWP. Any arbitrary polarization state can be generated by properly orienting these wave plates. The solid box represents a polarization state analyzer and includes a polarizer (P) and an optional QWP.

We know that in quantum systems, the analyzer measurements can be computed using Born's rule [1], which was outlined in chapter 3. A beam of single photons identically prepared in the state $|\psi\rangle$ is subjected to a linear polarizer oriented at α (labeled P in figure 5.6). The probability of photodetection is then given by (the QWP in the analyzer is not yet inserted)

$$
\begin{aligned}
P_{|\alpha\rangle} &= |\langle \alpha | \psi \rangle|^2 \\
&= |(\cos \alpha \langle H| + \sin \alpha \langle V|)(\cos \gamma |H\rangle + \sin \gamma e^{i\phi} |V\rangle)|^2 \\
&= \frac{1}{2}(1 + \cos(2\gamma)\cos(2\alpha) + \sin(2\gamma)\cos(\phi)\sin(2\alpha)),
\end{aligned}
\tag{5.41}
$$

which clearly resembles a Fourier series in α. Simple calculations show that the coefficients in this Fourier series are indeed the Stokes parameters, similar to what led to equations (2.27), (2.28), and (2.29). The Stokes parameters are defined through

$$
\begin{aligned}
S_0 &= P_{|H\rangle} + P_{|V\rangle}, \\
S_1 &= P_{|D\rangle} - P_{|A\rangle}, \\
S_2 &= P_{|L\rangle} - P_{|R\rangle}, \quad \text{and} \\
S_3 &= P_{|H\rangle} - P_{|V\rangle},
\end{aligned}
\tag{5.42}
$$

which is just equation (2.18) seen in another light (probabilities are proportional to intensities). We can calculate these probabilities for the state $|\psi\rangle$ spelled out in equation (5.40). For example, $S_1 = P_{|D\rangle} - P_{|A\rangle}$ can be computed in the following manner,

$$
P_{|D\rangle} = P(|D\rangle || \psi\rangle)
\tag{5.43}
$$

$$
= \left| \frac{1}{\sqrt{2}}(\langle H| + \langle V|)(\cos \gamma |H\rangle + e^{i\phi} \sin \gamma |V\rangle) \right|^2
\tag{5.44}
$$

$$
= \frac{1}{2}(1 + \sin(2\gamma)\cos \phi).
\tag{5.45}
$$

Similarly,

$$
P_{|A\rangle} = P(|A\rangle || \psi\rangle)
\tag{5.46}
$$

$$
= \left| \frac{1}{\sqrt{2}}(\langle H| - \langle V|)(\cos \gamma |H\rangle + e^{i\phi} \sin \gamma |V\rangle) \right|^2
\tag{5.47}
$$

$$
= \frac{1}{2}(1 - \sin(2\gamma)\cos \phi).
\tag{5.48}
$$

Hence,

$$
P_{|D\rangle} - P_{|A\rangle} = \sin(2\gamma)\cos \phi
\tag{5.49}
$$

indicating that the coefficient of $\sin(2\alpha)$ term in equation (5.41) is indeed the Stokes parameter S_1. Similarly, $\cos(2\gamma)$, the coefficient of $\cos(2\alpha)$ is found to be equal to the Stokes parameter S_3. This identifies the Fourier coefficients and Stokes parameters, allowing one to write, identically to equation (2.25),

$$I(\alpha) = \frac{1}{2}(S_0 + S_3\cos(2\alpha) + S_1\sin(2\alpha)). \tag{5.50}$$

Although equation (5.41) is for single photons while equation (5.50) is originally described for coherent light, it has been shown that the coefficients are indeed expressions for the respective Stokes parameters [4, 12], which describe partially coherent classical light as well. Similarly, equation (5.41) is also expressible in terms of the parameters of the polarization ellipse [13] in the following manner:

$$P_{|\alpha\rangle} = \frac{1}{2}(1 + \cos(2\chi)\cos(2\psi)\cos(2\alpha) + \cos(2\chi)\sin(2\psi)\sin(2\alpha)). \tag{5.51}$$

The Stokes parameters characterize the unknown polarization state of the light leaving the state generator. Therefore, if we have the probability measurements at a number of analyzer orientations, then we can make a Fourier curve fit [12, 14] to find the three Stokes parameters S_0, S_3 and S_1.

Now, let us analyze how the hippopede method bypasses the need for Fourier curve fitting. From equation (2.19), we know that knowledge of the two angles (χ, ψ) associated with the polarization ellipse allows us to immediately calculate the Stokes parameters $(S_3$ and $S_1)$. Moreover, if we know the sign of χ, we can also uniquely determine the value of S_2 [12].

We can vary the analyzer angle α and plot the graph between $\sqrt{P_{|\alpha\rangle}}\cos\alpha$ and $\sqrt{P_{|\alpha\rangle}}\sin\alpha$ to obtain the hippopedal pattern. It is evident from figure 5.5 that we can find the angular parameters of the corresponding ellipse as it sits inside the hippopede, i.e., we can directly measure the orientation angle ψ from the hippopedal figure and can calculate the ellipticity angle χ using the AR of the hippopede given by

$$\chi = \pm\tan^{-1}\left(\frac{q}{p}\right), \tag{5.52}$$

where q and p denote the lengths of the semiminor and semimajor axes of the hippopede. Subsequently, these two angles (χ, ψ) specify the Stokes parameters $(S_3$ and $S_1)$ deduced from equation (2.19). However, there still remains ambiguity about the handedness of the polarization, which in reality corresponds to the sign of S_2.

For this purpose, we must perform the optical version of the circularly polarized antenna method. For this specific purpose, we can place a QWP before the polarizer to find the handedness of the polarization (figure 5.6). We orient the analyzer QWP at $\pm 45°$ and the polarizer at $0°$ with respect to the horizontal. The probability expression then reduces to

$$P_{|L\rangle/|R\rangle} = \frac{1}{2}(1 \mp \sin(2\chi)). \tag{5.53}$$

This test of helicity is based on two projective measurements. If the probability of detecting $|L\rangle$ photons is larger than the probability of detecting $|R\rangle$ photons $P_{|L\rangle} > P_{|R\rangle}$, then the polarization is left-handed. On the other hand, $P_{|L\rangle} < P_{|R\rangle}$, the polarization is right-handed. If the unknown polarization is linearly polarized, we obtain $P_{|L\rangle} = P_{|R\rangle}$.

Although we have derived these expressions for a beam of single photons, similar expressions can be derived for coherent light, where we will get intensities instead of probabilities. We can test this classical versus quantum analogy for a single beam of light by comparing results with and without coincidence detection, corresponding to heralded single photons or classical light. Let us see what these experiments look like.

5.2.3 The experiment

The experimental setup, as depicted schematically in figure 5.6 and photographically in figure 5.7, is based on that of experiment Q1, with the addition of a few optical elements. As in the previous experiments, the detection of the photons striking detector A projects photons of the beam directed at detector B into single-photon states. Using the state-generation wave plate (QWP or HWP), we generate a beam of single photons with different polarization profiles and use the polarization pattern method for estimating the polarization-encoded quantum states. We compare the

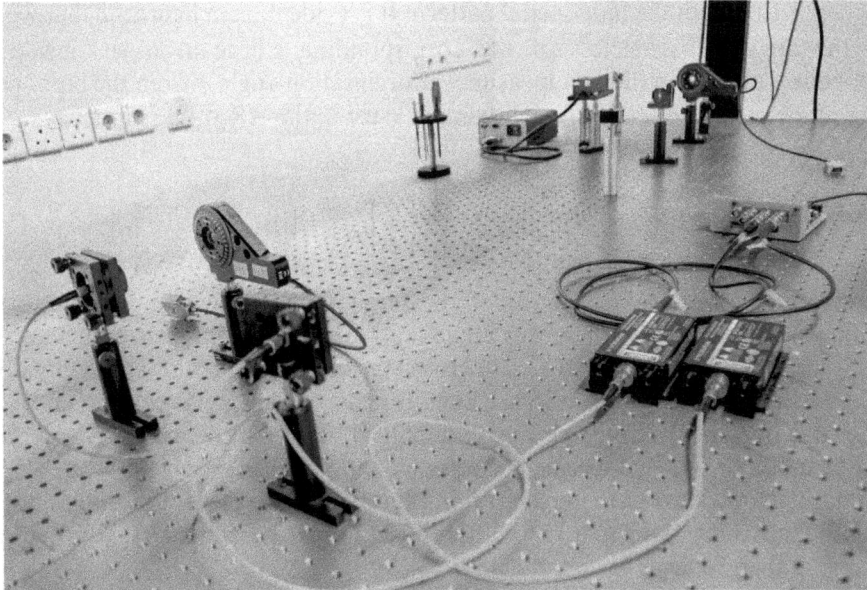

Figure 5.7. Photograph of the polarization pattern measurement setup, which is experiment Q4.

results of the experiment with those of experiment Q3. Finally, we compute the Stokes parameters from the probability differences mentioned in equation (5.42).

Simulated and experimental hippopedal plots for the degenerate polarization states are illustrated in figure 5.8. Making the aforementioned geometric measurements, we determine ellipse parameters from the plots and also compute Stokes parameters for various input polarization states. The results of this quantum state estimation process are listed in table 5.4. We have performed this analysis for both single counts (beam B considered alone) and coincidence counts (beam B conditioned on beam A). The former corresponds to a classical beam of light, while the latter corresponds to single-photon states. If we compare the Stokes parameters of the quantum (coincidence counts) and classical (single counts) versions of light, then we observe that the polarization properties are similar in both descriptions of light. Despite this similarity, it must be kept in mind that the statistics of photodetection are very different for the quantum and classical cases, as demonstrated by the $g^{(2)}(\tau)$ experiments performed in experiment Q2.

The experimental results confirm that we can draw an interesting analogy between optical polarimetry and polarization profiling of antennas. According to this analogy, one can map the optical state-generating system to the transmitting antenna, and likewise can map the optical analyzer to the reception antenna. Moreover, the results of this polarimetry experiment agree with those of the quantum state estimation scheme of experiment Q3.

Another interesting aspect is the remarkable agreement between the single-photon and coincidence-photon results. Although the results shown in chapter 4, based on the absence of coincidence detections of single photons, validate a quantum picture distinct from the classical realm, the correspondence highlighted in this section is a convincing manifestation of the similarity that exists between classical and quantum regimes of light. The classical picture of light corresponds to coherent light beams, such as those of lasers. In our experiment, the beams detected by individual detectors correspond to low-intensity classical beams. The quantum picture of light, of course, corresponds to heralded single photons. This experiment and the related discussion can be employed as interesting pedagogical arguments in the laboratory.

A benefit of the technique discussed in this experiment is that one can perform a few geometric measurements and determine the unknown polarization state from the polar graphs. The hippopedal plots are generated by recording photocounts with respect to various orientations of the analyzing polarizer. The geometric measurements fetch us the angular parameters of the ellipse, and by using these parameters we can compute the three Stokes parameters S_0, S_1, and S_3. However, to uniquely determine the handedness or helicity of the state, we need to place a QWP in the setup and make just two additional measurements.

So far, we have ensured that the photons produced by our source are truly quantum in nature and that we can manipulate their polarization when required. The polarization is also directly measurable and can be visualized. In the next chapter, we will exploit signature quantum properties investigating entanglement and nonlocality, phenomena that are unique to the quantum world.

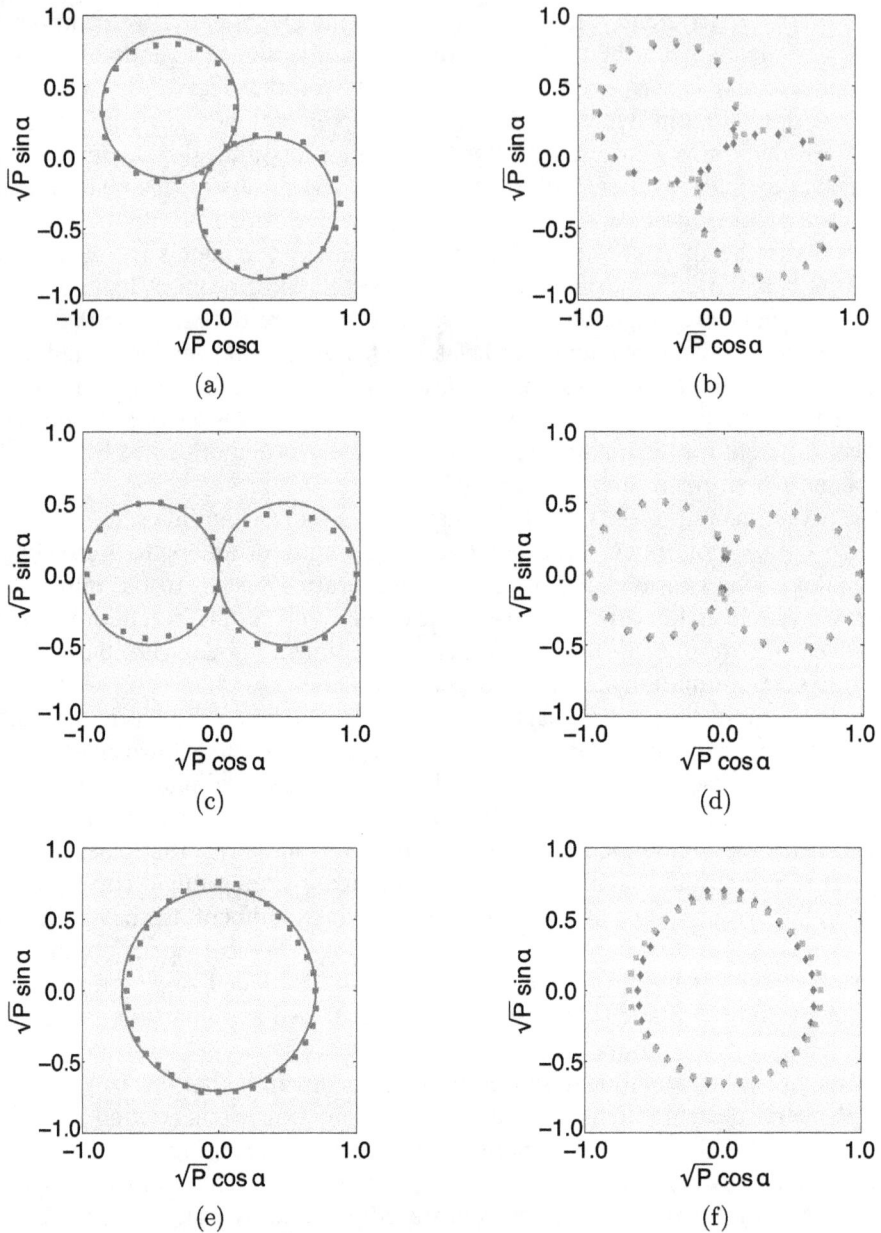

Figure 5.8. Polarization patterns for canonical polarization states. The generated polarization states include anti-diagonal (a, b), horizontal (c, d), and left-hand circular (e, f). (a), (c), and (e) compare simulated plots (smooth curves) with experimental data (discrete points), while (b), (d), and (f) compare the polarization state of classical (coherent) light with quantum (single-photon) light. The red points represent measurements for a classical light beam, while the blue points represent measurements corresponding to single photons. Adapted with permission from [6]. © The Optical Society.

Table 5.4. Experimental results of quantum state estimation and optical polarimetry for canonical polarization states. The column labeled 'quantum state estimation' quotes the Stokes parameters obtained using the technique discussed in experiment Q3. The columns labeled 'quantum hippopede' and 'classical hippopede' enlist the Stokes parameters obtained from the polarization patterns (shown in figure 5.8) using the technique developed in experiment Q4. The quantum hippopede is based on coincidence counts accounting for single photons, whereas the classical hippopede is based on individual detector counts accounting for a coherent light beam.

Polarization state	Quantum state estimation	Quantum hippopede	Classical hippopede
$|H\rangle$	$S_0 = 1.00 \pm 0.00$	1.00 ± 0.04	1.00 ± 0.04
	$S_3 = 0.97 \pm 0.00$	0.95 ± 0.04	0.92 ± 0.04
	$S_1 = -0.06 \pm 0.00$	-0.16 ± 0.04	-0.15 ± 0.04
$|V\rangle$	$S_0 = 1.00 \pm 0.00$	1.00 ± 0.02	1.00 ± 0.02
	$S_3 = -0.87 \pm 0.00$	-0.98 ± 0.02	-0.92 ± 0.02
	$S_1 = 0.21 \pm 0.00$	-0.08 ± 0.02	-0.09 ± 0.02
$|D\rangle$	$S_0 = 1.00 \pm 0.00$	1.00 ± 0.00	1.00 ± 0.00
	$S_3 = 0.12 \pm 0.00$	0.04 ± 0.02	0.04 ± 0.02
	$S_1 = 0.69 \pm 0.00$	1.00 ± 0.00	1.00 ± 0.00
$|A\rangle$	$S_0 = 1.00 \pm 0.02$	1.00 ± 0.02	1.00 ± 0.03
	$S_3 = -0.02 \pm 0.00$	0.11 ± 0.02	0.13 ± 0.03
	$S_1 = -0.90 \pm 0.00$	-0.97 ± 0.02	-0.93 ± 0.03
$|L\rangle$	$S_0 = 1.00 \pm 0.00$	1.00 ± 0.04	1.00 ± 0.02
	$S_3 = 0.07 \pm 0.00$	-0.04 ± 0.04	0.06 ± 0.02
	$S_1 = 0.08 \pm 0.00$	-0.02 ± 0.04	-0.03 ± 0.02
$|R\rangle$	$S_0 = 1.00 \pm 0.00$	1.00 ± 0.04	1.00 ± 0.02
	$S_3 = -0.12 \pm 0.00$	-0.07 ± 0.04	0.06 ± 0.02
	$S_1 = 0.03 \pm 0.00$	-0.04 ± 0.06	-0.04 ± 0.02

References

[1] Beck M 2012 *Quantum Mechanics: Theory and Experiment* (Oxford: Oxford University Press)

[2] Lvovsky A I 2018 *Quantum Physics: An Introduction Based on Photons* (Berlin: Springer)

[3] Jones R C 1941 A new calculus for the treatment of optical systems description and discussion of the calculus *J. Opt. Soc. Am.* **31** 488–93

[4] Collett E 2005 *Field Guide to Polarization* (Bellingham, WA: SPIE)

[5] Wolf E 1954 Optics in terms of observable quantities *Il Nuovo Cimento (1943–1954)* **12** 884–8

[6] Waseem M H, Anwar M S *et al* 2019 Hippopedal intensity plots: drawing comparisons between antenna and optical polarimetry *Appl. Opt.* **58** 8442–8

[7] Mayes T W 1976 Polar intensity profile of elliptically polarized light *Am. J. Phys.* **44** 1101–3

[8] Stutzman W L and Thiele G A 2013 *Antenna Theory and Design* (Hoboken, NJ: Wiley)

[9] Lawrence J D 2013 *A Catalog of Special Plane Curves* (New York, NY: Dover Publications)

[10] Moroni L 2017 The toric sections: a simple introduction *arXiv preprint* arXiv:1708.00803

[11] Bohnert J I 1951 Techniques for handling elliptically polarized waves with special reference to antennas: part IV—measurements on elliptically polarized antennas *Proc. IRE* **39** 549–52

[12] Goldstein D H 2016 *Polarized Light* (Boca Raton, FL: CRC Press)

[13] Collett E and Schaefer B 2008 Visualization and calculation of polarized light. I. The polarization ellipse, the Poincaré sphere and the hybrid polarization sphere *Appl. Opt.* **47** 4009–16

[14] Berry H G, Gabrielse G and Livingston A E 1977 Measurement of the Stokes parameters of light *Appl. Opt.* **16** 3200–5

IOP Publishing

Quantum Mechanics in the Single-Photon Laboratory
(Second Edition)

Muhammad Sabieh Anwar, Faizan-e-Ilahi, Syed Bilal Hyder and Muhammad Hamza Waseem

Chapter 6

Entanglement and nonlocality

In the previous chapters, our discussion about quantum behavior was confined to the signal beam, while the idler beam only acted as a conditioner to ascertain the detection of single photons in the signal beam. From this chapter onward, we will expand our discussion and investigate the *joint* quantum properties of the idler and signal beams, i.e., the joint polarization description of the two-photon system. Per previous practice, we define the following naming scheme to conveniently describe these paired photon experiments.

- NL1: Freedman's test of locality
- NL2: The Clauser–Horne–Shimony–Holt (CHSH) test of locality
- NL3: Hardy's test of locality

These experiments are litmus tests for entanglement. As a consequence of entanglement [1] between at least two particles, predictions of locality are violated. Let us start by briefly discussing concepts of entanglement and locality.

6.1 Entanglement and nonlocality: a survey

Physical systems involving two or more particles can demonstrate certain behaviors that have no classical counterpart. In particular, the states of two particles, photons in our case, can become 'entangled'. As discussed in chapter 3, two or more particles are said to be entangled if their combined state cannot be written as a product of individual particle states. In other words, these states are inseparable.

Over the last few decades, apart from their central role in discussions of nonlocal quantum correlations (the subject of this chapter) [2], entangled particles have been exploited for applications such as quantum cryptography [3, 4], quantum teleportation [5], dense coding [6], and quantum computing [7]. The 2022 Nobel Prize in Physics was awarded to Alain Aspect, John F. Clauser, and Anton Zeilinger for

doi:10.1088/978-0-7503-6315-0ch6

their investigation of quantum entanglement and nonlocality, and the many technological advances that these concepts offer.

Entangled photons can show that quantum mechanics violates predictions of 'locality', sometimes also referred to as 'local realism'. Locality means that measuring one particle's state cannot instantaneously impact another particle's state. Reality requires that all physical systems possess definite values, whether we measure them or not [2]. Hence, local realism dictates that if we have a source producing pairs of photons, then the state of each photon in a pair is defined as soon as it leaves the source and that any measurement that can be performed on one photon of the pair cannot instantaneously affect the state of the other pair member. Therefore, all the properties of the pair are encoded into and distributed to the pair at the time of generation.

As per conventions of the scientific community, we will refer to local realism simply as *locality*. The assumption of locality is normally taken for granted because all classical systems obey it. But this is not necessarily true for quantum systems. We are going to develop a series of experiments to demonstrate that quantum systems violate the assumption of locality.

The idea of entanglement was popularized in the physics community through a *gedankenexperiment* (thought experiment) by Einstein, Podolsky, and Rosen (EPR), published in 1935 [2]. Entangled photons, when individually seen, may seem to be randomly polarized. However, the polarization states, when seen collectively as a composite bipartite system, may show strong correlations which cannot be described by any classical theory. The 'Copenhagen interpretation' [8], championed by Bohr and his followers, claims that such correlations arise from the nonlocality associated with measurement—measuring the state of one particle collapses the entangled state of both particles instantly. Contrary to Bohr, the EPR trio in their paper argued that such 'action at a distance' (nonlocality) is impossible, and they claimed that quantum theory was incomplete because it failed to furnish a complete description of reality [2].

For many years, this paradox was termed a philosophical debate. Physicists followed the 'shut up and calculate' approach to quantum mechanics, which worked remarkably well in practical situations. It was John Bell who showed that locality could, in fact, be experimentally tested. In 1964, he derived an inequality that must be obeyed by any theory based on locality [9, 10]. Bell's original inequality was for an idealized system. Later, other researchers derived similar but testable inequalities [11–13]. These are collectively called Bell's inequalities. These inequalities have been violated experimentally, proving that entangled systems demonstrate nonlocality [13–15].

In 1972, the first Bell's inequality was tested [13]; this rather elementary inequality was derived by Freedman [13]. Bell's inequality typically tested in optical systems is the CHSH inequality [11, 16, 17]. Greenberger, Horne, and Zeilinger (GHZ) went another step ahead and showed that, instead of violating an inequality, an 'all or nothing' test of locality [18, 19] could be performed. This is a stricter test of nonlocality in which a local theory predicts a certain result, whereas quantum theory predicts a totally opposite result. This experiment was significantly more difficult and involved three entangled particles instead of two, as required in other tests. Eventually, the experiment was performed, and its results also agreed with quantum mechanical predictions [14]. In 1993, Hardy derived yet another version of Bell's theorem [12, 20], which is comparatively easier to comprehend than the CHSH test. Experiments based on Hardy's proposal have also been performed, and their outcomes, yet again, confirm the theoretical predictions of quantum mechanics [15, 21, 22].

In this chapter, we will use entangled photon-pair experiments to study and perform three tests of locality based on the ideas of Freedman [13, 23], Hardy [12, 24], and CHSH [11, 16].

6.2 The proverbial Alice and Bob experiment

Many analogies have been employed in the literature to describe entanglement experiments for testing locality [25–27]. We will present a description, closely following that of reference [24], which corresponds to our single-photon polarization-based systems.

Consider a source that produces pairs of photons. The two photons in each pair move in different directions: one travels to a party named Alice, while the other travels to Bob; see figure 6.1. The source is exactly midway between the two parties, meaning they receive the photons simultaneously. For the sake of simplicity, let us assume that the photons are fired toward the two parties at regular intervals. Each party has a linear polarizer (P) and a detector (labeled A for Alice and B for Bob). Suppose that Bob aligns his polarizer along a particular angle, and his detector picks a photon. In that case, he can say that the detected photon was polarized at an angle parallel to his polarizer axis. In other words, Bob has measured a photon with this polarization. A similar explanation goes for Alice.

Alice and Bob decide to perform the experiment. As the photon pairs are fired at regular intervals, Alice and Bob perform measurements. Alice randomly orients her polarizer along the angle θ_{Ai} and Bob randomly orients his polarizer along θ_{Bi}. There is a vast distance between Alice and Bob, and they do not communicate while performing the measurements. Once all the measurements have been performed, the two parties meet to check for correlations between their measured photocounts.

Alice and Bob find that the measured photons when individually seen, seem to be randomly polarized; yet they show unexpectedly strong correlations when the angles θ_{Ai} and θ_{Bi} coincide. This correlation of individually random outcomes cannot be

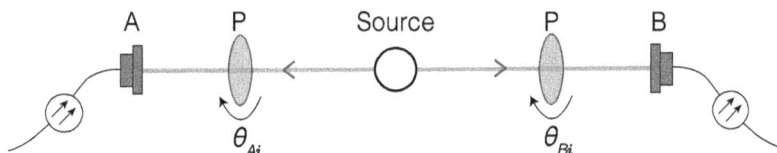

Figure 6.1. The EPR thought experiment in terms of polarization-entangled photons. A source sends out pairs of photons in opposite directions, toward Alice and Bob, who randomly orient their polarizers at angles θ_{Ai} and θ_{Bi} respectively, and count the photons detected at detectors A and B, respectively.

described by any local model (when seen collectively). This is, in fact, an optical incarnation of the famous EPR thought experiment. Let us actualize this virtual experiment and then look at it through the lens of Freedman [13], Hardy [12], and CHSH [11].

6.3 Generating polarization-entangled photons

Entangled states are an essential ingredient of quantum physics and have no classical counterpart [1]. Nowadays, the SPDC process [22, 28, 29] offers a popular and accessible source of entangled states.

To understand the polarization-entanglement of photons, let us revisit spontaneous parametric downconversion (SPDC). In experiments Q1–Q4, we were not concerned about the polarization state of idler photons because each idler photon was merely used to herald the detection of the corresponding signal photon. For nonlocality experiments, on the contrary, we are required to consider the complete polarization state of the photon pair.

We use coincidence detection to ensure *single*-photon detections in which each detector click corresponds to the detection of one photon. Therefore, when we refer to a two-photon system, we mean an experimental system involving two beams of *single* photons, which is automatically ensured when they are detected coincidentally.

The polarization-entangled two-photon state is generated using the method proposed by Kwiat [28], in which two type-I downconversion crystals with mutually orthogonal optical axes are used. One crystal downconverts the vertically polarized pump photons into horizontally polarized ones, while the other is responsible for the downconversion of horizontally polarized pump photons into vertically polarized photon pairs. In experiments Q1–Q4, only one of these crystals was effectively used because the pump beam was vertically polarized. However, if the pump beam has both the horizontal and vertical polarization components, both of the crystals will downconvert the incoming light (see figure 6.2). The downconverted beam produced by the two crystals is expressible as the superposition

$$A|H\rangle_A|H\rangle_B + Be^{i\phi}|V\rangle_A|V\rangle_B. \tag{6.1}$$

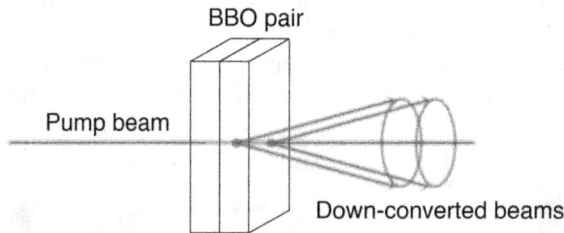

BBO pair

Pump beam

Down-converted beams

Figure 6.2. Spontaneous parametric downconversion to generate entangled photon pairs. One crystal downconverts horizontally polarized photons of the pump beam into vertically polarized photon pairs, while the other downconverts vertically polarized photons into horizontally polarized photon pairs. When the two crystals are sandwiched, and the pump beam is in a superposition of horizontally and vertically polarized photons, an entangled state of two photons is generated.

Here, subscript A corresponds to the photon going toward Alice, whereas subscript B corresponds to the photon going toward Bob. Furthermore, due to a finite, nonzero width of the BBO crystals, one of the components gets downconverted earlier than the other, which introduces the phase difference φ. This phase is encapsulated by $e^{i\phi}$ in equation (6.1). For the sake of brevity, we can also represent this state as

$$A|HH\rangle + Be^{i\phi}|VV\rangle \tag{6.2}$$

and just keep in mind that the first member of each ket represents the polarization of Alice's photon, while the second member represents that of Bob's photon.

Any desired state with the required A, B, and φ can be produced by altering the polarization state of the pump beam. This is done by placing a half-wave plate (HWP) and a birefringent quartz plate in the path of the pump beam. The coefficients A and B can be controlled by changing the orientation of the HWP, while φ can be compensated for by introducing a quartz plate in the pump beam's path. We can generate arbitrary linear combinations of the states $|HH\rangle$ and $|VV\rangle$ using this arrangement.

We will find that to explain outcomes of experiments performed using the downconverted beams of figure 6.2, we need to interpret the state (6.2) as that of photons being in a superposition of $|HH\rangle$ and $|VV\rangle$ states. The state (6.2) is pure. In practice, however, a mixed state is produced, which could be an incoherent mixture of the state (6.1) combined with other states. Such a state is described by a density matrix rather than a ket. The quantum state tomography (QST) experiment, the subject of chapter 8, allows us to estimate this density matrix.

For an entangled state, the polarization measurements may seem purely random when individually analyzed for the two beams, yet they are strongly correlated when seen collectively. Furthermore, neither of the photons has a well-defined, objectively articulated polarization *before* the measurement. However, if we make a measurement and determine the polarization of either of the photons, then the other photon's polarization is automatically determined. This fact also holds for measurements performed in other bases of measurement [30]. Therefore, whenever the analysis angles match up ($\theta_{Ai} - \theta_{Bi} = 0°$ or $90°$), no matter what they are in the absolute sense, the results are correlated. No theory of classical physics can account for such a strong correlation. Let us discuss the experimental procedure of generating entangled states before delving into the details of the nonlocality experiments one by one.

6.3.1 Experimental setup

All of the experiments in this section will employ a very similar four-detector setup. If we align the complete setup once, then we can perform all the nonlocality experiments covered in this chapter. Hence, this description is universal. We have learned several alignment techniques while progressing through the previous chapters; these techniques are also applicable here. Figure 6.3 is a combined schematic view of the setup for NL1, NL2, and NL3.

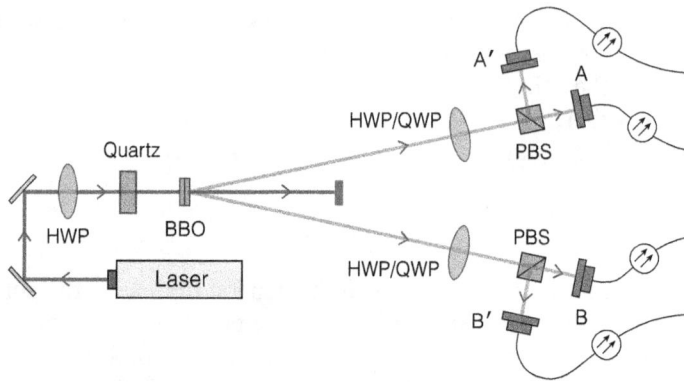

Figure 6.3. A four-detector setup containing two polarizing beam splitters and four detectors (A, A', B, and B'). We can generate many entangled states by varying the orientation of the wave plates and the quartz plate in the pump beam's path. The measurement basis is controlled by HWP and/or quarter-wave plates placed in the paths of the downconverted beams.

The two polarizing beam splitters (PBSs) can be aligned using the double back-propagation laser technique that we learned in experiment Q2. We use the signal and idler beam wave plates to change the measurement basis, just as we did in experiment Q3.

6.3.2 Measuring probabilities with four detectors

Considering the four-detector measurement setup in figure 6.3, the idler photons move toward Alice and fall on either of her detectors (A and A'). On the other side, the signal photons move toward Bob and fall on either of his detectors (B and B'). We want to measure the rate of events when Alice's and Bob's detectors register the photons simultaneously. Therefore, we are interested in counting coincidence events in a certain time period. For instance, $N_{AB'}$ denotes the number of coincidences occurring at the detectors A and B' in a certain counting period.

Once we have measured the coincidence counts represented by N_{AB}, $N_{AB'}$, $N_{A'B}$, and $N_{A'B'}$, we can compute the probability that the photons detected by Alice and Bob had certain polarizations. For instance, let us say that Alice's analysis angle is θ_A. Photons polarized along the analysis angle θ_A will be transmitted by the PBS toward detector A, and ones orthogonal to it, $\theta_{A'}$ will be reflected toward A'. Similarly, if Bob's analysis angle is θ_B, the photons polarized along the analysis angle θ_B will be transmitted to detector B, while those perpendicular to it, $\theta_{B'}$, will be reflected to detector B'. With this convention set, we define the joint probability of the four heralded detections as

$$P(\theta_A, \theta_B) = \frac{N_{AB}}{N_{AB} + N_{AB'} + N_{A'B} + N_{A'B'}}, \tag{6.3}$$

$$P(\theta_A, \theta_{B'}) = \frac{N_{AB'}}{N_{AB} + N_{AB'} + N_{A'B} + N_{A'B'}}, \tag{6.4}$$

$$P(\theta_{A'}, \theta_B) = \frac{N_{A'B}}{N_{AB} + N_{AB'} + N_{A'B} + N_{A'B'}}, \tag{6.5}$$

$$P(\theta_{A'}, \theta_{B'}) = \frac{N_{A'B'}}{N_{AB} + N_{AB'} + N_{A'B} + N_{A'B'}}. \tag{6.6}$$

From now on, we will represent probabilities in the form $P(\theta_A, \theta_B)$. Where θ_A is Alice's analysis angle, and θ_B is Bob's analysis angle.

We saw in experiment Q1 that turning the pump HWP by 45° turns the input polarization from $|V\rangle$ to $|H\rangle$, and the other BBO starts downconverting the incoming light. However, if we turn the HWP by an angle between 0° and 45°, the input light will contain both $|V\rangle$ and $|H\rangle$ polarized light, and both of the BBO crystals will participate in downconversion. We now generate the maximally entangled states, also known as the Bell states.

6.3.3 Generating Bell states

The four Bell states are defined as

$$|\Phi^+\rangle = \frac{1}{\sqrt{2}}(|HH\rangle + |VV\rangle), \tag{6.7}$$

$$|\Phi^-\rangle = \frac{1}{\sqrt{2}}(|HH\rangle - |VV\rangle), \tag{6.8}$$

$$|\Psi^+\rangle = \frac{1}{\sqrt{2}}(|HV\rangle + |VH\rangle), \quad \text{and} \tag{6.9}$$

$$|\Psi^-\rangle = \frac{1}{\sqrt{2}}(|HV\rangle - |VH\rangle). \tag{6.10}$$

Using type-I downconversion, we can generate only two of these states, $|\Phi^+\rangle$ and $|\Phi^-\rangle$, because type-I downconversion generates photon pairs with identical polarization $|HH\rangle$ and $|VV\rangle$. To generate $|\Psi^+\rangle$ and $|\Psi^-\rangle$ states, one would require type-II downconversion crystals.

As part of these experiments, we will need to switch between measurement bases. Table 6.1 shows what states the four detectors will measure when aligned along the different measurement bases.

Table 6.1. What the detectors measure when they register clicks.

Measurement basis	AB	A'B	AB'	A'B'
$\{|H\rangle, |V\rangle\}$	$|VV\rangle$	$|HV\rangle$	$|VH\rangle$	$|HH\rangle$
$\{|D\rangle, |A\rangle\}$	$|AA\rangle$	$|DA\rangle$	$|AD\rangle$	$|DD\rangle$
$\{|L\rangle, |R\rangle\}$	$|RR\rangle$	$|LR\rangle$	$|RL\rangle$	$|LL\rangle$

Both $|\Phi^+\rangle$ and $|\Phi^-\rangle$ have equal superposition of $|HH\rangle$ and $|VV\rangle$ states. If we turn the pump beam HWP by 22.5°, the input to the BBO stack will contain equal portions of photons in $|H\rangle$ and $|V\rangle$ states. Hence, the downconverted beam will have equal amplitudes of $|VV\rangle$ and $|HH\rangle$ pairs as shown by the equation

$$\frac{1}{\sqrt{2}}(|V\rangle + |H\rangle) \xrightarrow{BBO} \frac{1}{\sqrt{2}}(|HH\rangle + e^{i\phi}|VV\rangle) = |\psi\rangle. \qquad (6.11)$$

At this point, referring to table 6.1, if we turn the measurement basis to $\{|H\rangle, |V\rangle\}$ in both channels, then the AB and A' B ' counts should be comparable.

Notice that although the two portions of the state have been downconverted and the state looks like a Bell state, we are not quite there yet. There is a relative phase φ between the two components. This phase originates because the two BBO crystals have finite, nonzero lengths and a finite distance between them. While passing through the BBO stack, one component of the input polarization gets down-converted a bit later than the other, and that shows up as the phase φ. To mitigate this phase, a compensatory phase must be introduced in the pump beam. We do this by placing a plate made of a birefringent material, quartz, right after the pump beam HWP. The quartz plate affixed in its tiltable mount is shown in figure 6.4a. By placing it in the correct orientation, we can introduce a relative phase between the two components of light. With this quartz plate in the setup, we now have two phase-inducing components. The phase by one of these components (the BBO) is fixed, while we can change the phase introduced by the other (the quartz).

Note that for the Bell states $|\Phi^{\pm}\rangle$, we have $\phi = 0$ or π. To prepare $|\Phi^+\rangle$, we bring the phase as close to zero as possible by performing a simple test. We turn the measurement basis to $\{|D\rangle, |A\rangle\}$ in both channels. The counts $A'B$ and AB' now

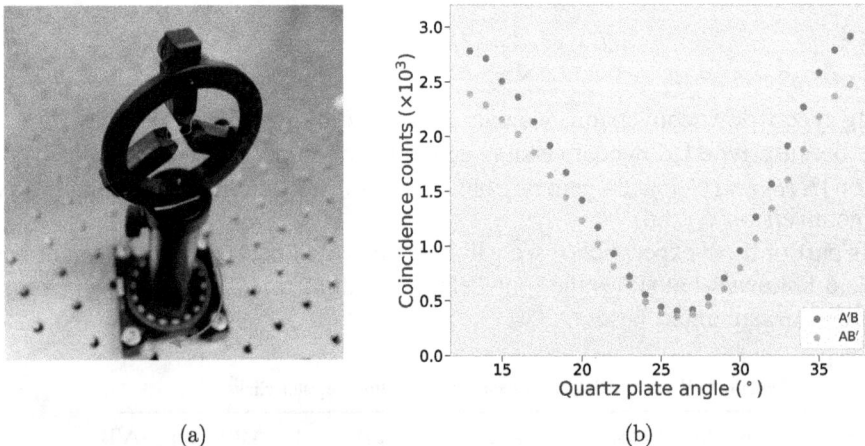

(a)   (b)

Figure 6.4. (a) Quartz plate held in a mount with a rotatable base. By rotating the mount (as shown by the curved arrow), we can change the tilt of the plate and, hence, the phase introduced by it. (b) The quartz plate is tilted such that the cross counts $A'B$ and AB' are at their minimal level. In this graph, the minimum cross counts correspond to the tilt angle of 27°.

account for the measurements in state $|DA\rangle$ and $|AD\rangle$, respectively. When the phase ϕ changes, the counts will also periodically change, as can be seen here,

$$
\begin{aligned}
|\langle AD|\psi\rangle|^2 = |\langle DA|\psi\rangle|^2 &= \left| \frac{1}{2\sqrt{2}}(1 - e^{i\phi}) \right|^2 \\
&= \frac{1}{8}(1 - e^{i\phi})(1 - e^{-i\phi}) = \frac{1}{4}\sin^2\left(\frac{\phi}{2}\right)
\end{aligned}
\tag{6.12}
$$

For the Bell state $|\Phi^+\rangle$, we require $\phi = 0$; which means that the value of $|\langle DA|\psi\rangle|^2$ should approach zero. Therefore, in order to prepare a state as close as possible to $|\Phi^+\rangle$, we need to change the tilt of quartz crystal until the cross counts are minimized, as shown in figure 6.4b. Following a similar reasoning, the phase term needs to be -1 for state $|\Phi^-\rangle$, so we need to change the tilt of the quartz crystal until the cross counts maximize. With the phase dealt with, we now have the (theoretically) perfect Bell state. All of the tests for nonlocality in this chapter will require one of these Bell states except for Hardy's test. Hardy's test requires a different input state, which will be discussed later.

6.4 NL1: Freedman's test of locality

Freedman's test is historically significant because it was the first experimental test of any Bell's inequality [13]. It was also the focus of the Nobel Laureate, Alain Aspect's paper [31] published in 1981. Although the newer tests serve as a great exercise for an undergraduate laboratory, Freedman's test may be presented earlier to bring home the basic ideas of the Bell tests.

Normally, Freedman's test is performed using two detectors instead of a four-detector setup like ours. To keep the discussion simple and close to the original, we will use the two-detector approach while deriving the inequality. Obviously the setup *with four detectors* can easily be used in the two-detector configuration, by not processing the photocounts A' and B', which are nevertheless available to us.

6.4.1 Freedman's inequality

We reproduce the derivation of Freedman's inequality following reference [23]. This inequality is based on the assumptions of locality, reality, and hidden variables, which will be explained shortly. Consider six real numbers, namely x_1, x_2, y_1, y_2, X, and Y, that obey the following relations:

$$0 \leqslant x_1 \leqslant X, \tag{6.13}$$

$$0 \leqslant x_2 \leqslant X, \tag{6.14}$$

$$0 \leqslant y_1 \leqslant Y, \quad \text{and} \tag{6.15}$$

$$0 \leqslant y_2 \leqslant Y. \tag{6.16}$$

With these, we can define a term U such that

$$U \equiv x_1 y_1 - x_1 y_2 + x_2 y_1 + x_2 y_2 - Y x_2 - X y_1. \tag{6.17}$$

Considering equations (6.13)–(6.17), the following relation can be shown [23] to hold true:

$$-XY \leqslant U \leqslant 0. \tag{6.18}$$

Now, let us consider photon pairs produced from a source, as shown in figure 6.1. The two photons of any pair take different paths. We place a polarizer followed by a detector—let us call this assembly an analyzer—in each path. Let N_t denote the total number of photon pairs directed toward these analyzers. Let $N(\theta_A, \theta_B)$ represent the coincidence counts during the same duration, where θ_A and θ_B denote the analysis angles in Alice's path A and Bob's path B, respectively. If the measurement time period is sufficiently long, the probability of detection of the photon pairs is given by

$$P(\theta_A, \theta_B) = \frac{N(\theta_A, \theta_B)}{N_t}. \tag{6.19}$$

This probability cannot be directly measured due to nonideal detector efficiencies. To account for another factor contributing to the coincidence detections, we introduce a hidden variable λ. Keeping λ in consideration, the probability of detecting a photon pair becomes $P(\lambda, \theta_A, \theta_B)$. As per our current discussion, the hidden variable is the photodetection efficiency. Equation (6.19) then describes the average probability of detecting coincidence events for a number of photon pairs.

For a particular photon pair, let us consider a simple case where $P(\lambda, \theta_A, \theta_B)$ can either be 0 or 1. This implies that a coincidence event can either be detected or not detected and is predetermined (due to λ) with absolute certainty. It should be noted that the hidden variable λ lies in the confines of classical physics, and the probability it determines is well-defined regardless of whether any observer is aware of the value, i.e., λ obeys the reality assumption.

Practically speaking, however, the probability of coincidence detection $P(\lambda, \theta_A, \theta_B)$ may have values between 0 and 1 because there may be more than one hidden variables determining the detection probability for a particular event. If $p(\lambda)$ is used to denote the probability distribution of λ, we have

$$P(\theta_A, \theta_B) = \int P(\lambda, \theta_A, \theta_B) p(\lambda) \mathrm{d}\lambda. \tag{6.20}$$

If we take the locality assumption into account, a polarizer in the path of B has no effect on a photon in path A, and likewise a polarizer in path A has no effect on a photon in path B. We have

$$P(\lambda, \theta_A, \theta_B) = P(\lambda, \theta_A) P(\lambda, \theta_B). \tag{6.21}$$

The common hidden variable λ characterizes both the photons and lets us write the probability as a product of two measurement probabilities [23] in equation (6.21). The reason is that in local hidden variable theories, λ captures any prior information that is imbedded in the photon pairs, or the detection efficiencies, created at the time of the photon generation, that can lead to correlations in detections of the photon pairs.

We use $P(\lambda, \infty_A)$ to denote the detection probability of photon if no polarizer is placed in detector A's path. If we consider the no-enhancement assumption[1], we have

$$0 \leqslant P(\lambda, \theta_{A1}) \leqslant P(\lambda, \infty_A). \tag{6.22}$$

Similarly, for a different orientation of polarizer A, we have

$$0 \leqslant P(\lambda, \theta_{A2}) \leqslant P(\lambda, \infty_A). \tag{6.23}$$

Likewise, for polarizer B, we have

$$0 \leqslant P(\lambda, \theta_{B1}) \leqslant P(\lambda, \infty_B), \quad \text{and} \tag{6.24}$$

$$0 \leqslant P(\lambda, \theta_{B2}) \leqslant P(\lambda, \infty_B). \tag{6.25}$$

Therefore, comparing equations (6.13)–(6.16) with equations (6.22)–(6.25) and using equations (6.17) and (6.18), we can write

$$\begin{aligned}
- P(\lambda, \infty_{A1})P(\lambda, \infty_{B1}) &\leqslant P(\lambda, \theta_{A1})P(\lambda, \theta_{B1}) - P(\lambda, \theta_{A1})P(\lambda, \theta_{B2}) \\
&\quad + P(\lambda, \theta_{A2})P(\lambda, \theta_{B1}) + P(\lambda, \theta_{A2})P(\lambda, \theta_{B2}) \\
&\quad - P(\lambda, \theta_{A2})P(\lambda, \infty_B) - P(\lambda, \infty_A)P(\lambda, \theta_{B1}) \leqslant 0.
\end{aligned} \tag{6.26}$$

After multiplying by $p(\lambda)d\lambda$ and integrating, equation (6.26) becomes

$$\begin{aligned}
- P(\infty_A, \infty_B) &\leqslant P(\theta_{A1}, \theta_{B1}) - P(\theta_{A1}, \theta_{B2}) + P(\theta_{A2}, \theta_B1) \\
&\quad + P(\theta_{A2}, \theta_{B2}) - P(\theta_{A2}, \infty_{B1}) - P(\infty_A, \theta_{B1}) \leqslant 0.
\end{aligned} \tag{6.27}$$

At this point, we make another assumption, of rotational invariance, according to which the coincidence counts depend on the relative angle $\phi = |\theta_A - \theta_B|$ between the two analyzer settings and not on the individual angles, i.e., $P(\theta_A, \theta_B) = P(\phi)$. Hence, we can choose $\theta_{A1}, \theta_{A2}, \theta_{B1}$, and θ_{B2} according to the relation

$$|\theta_{A1} - \theta_{B1}| = |\theta_{A2} - \theta_{B2}| = |\theta_{A2} - \theta_{B1}| = \frac{|\theta_{A1} - \theta_{B2}|}{3} = \phi, \tag{6.28}$$

so that equation (6.27) becomes

$$-P(\infty_A, \infty_B) \leqslant 3P(\phi) - P(3\phi) - P(\theta_{A2}, \infty) - P(\infty_A, \theta_{B1}) \leqslant 0. \tag{6.29}$$

For a polarizer, it is well-known that $P(\phi) = P(\phi + 180°)$. Therefore, taking $P(202.5°) = P(22.5°)$ and $\phi = 67.5°$ in equation (6.29), we obtain

$$-P(\infty_A, \infty_B) \leqslant 3P(67.5°) - P(22.5°) - P(\theta_{A2}, \infty_B) - P(\infty_A, \theta_{B1}) \leqslant 0, \tag{6.30}$$

whereas plugging $\phi = 22.5°$ into equation (6.29) yields

$$-P(\infty_A, \infty_B) \leqslant 3P(22.5°) - P(67.5°) - P(\theta_{A2}, \infty_B) - P(\infty_A, \theta_{B1}) \leqslant 0. \tag{6.31}$$

---

[1] Adding a polarizer in the path of the light beam cannot increase the number of photodetections.

Subtracting equation (6.30) from equation (6.31) gives

$$-P(\infty_A, \infty_B) \leqslant 4P(22.5°) - 4P(67.5°) \leqslant P(\infty_A, \infty_B). \tag{6.32}$$

Multiplying equation (6.32) by N_t, we can rewrite it as

$$-N_0 \leqslant 4N(22.5°) - 4N(67.5°) \leqslant N_0. \tag{6.33}$$

Here, $N_0 = N(\infty, \infty)$ represents the number of coincidence measurements if there are no polarizers in the path of the two photon beams, and $N(\phi)$ denotes the number of coincidence detections when φ is the relative angle between the two polarizers. Finally, we can define a parameter δ and rewrite equation (6.33) as

$$\delta \equiv \left| \frac{N(22.5°) - N(67.5°)}{N_0} \right| - \frac{1}{4} \leqslant 0, \tag{6.34}$$

which is a quantity that can be directly determined from the measured photocounts. This inequality is known as Freedman's inequality.

If all assumptions made while deriving Freedman's inequality were valid, then $\delta \leqslant 0$ must hold. If experimental measurements violate this relation, then at least one of the assumptions is incorrect. The no-enhancement and rotational invariance assumptions can be conveniently tested in the laboratory [23, 32]. The remaining assumptions imply local hidden variables. They are ruled out if experiments confirm $\delta > 0$.

6.4.2 The quantum prediction for Freedman's test

The polarization-encoded Bell state we use for Freedman's test is represented as

$$|\Phi^+\rangle = \frac{1}{\sqrt{2}}(|HH\rangle + |VV\rangle). \tag{6.35}$$

If our source generates photons in the state $|\Phi^+\rangle$, the joint probability that Alice detects a photon polarized along angle θ_A and Bob detects a photon polarized along angle θ_B can be determined by Born's rule as

$$P_{\text{ideal}}(\theta_A, \theta_B) = |\langle\theta_A\theta_B||\Phi^+\rangle|^2 \tag{6.36}$$

if each polarizer is assumed to transmit 100% of the light polarized along its axis. This is a probability of coincidence detection. If the polarizer angles θ_A and θ_B are measured with respect to the horizontal axis of the lab frame, then we have

$$|\theta_A\rangle = \cos\theta_A|H\rangle + \sin\theta_A|V\rangle, \tag{6.37}$$

$$|\theta_B\rangle = \cos\theta_B|H\rangle + \sin\theta_B|V\rangle. \tag{6.38}$$

equation (6.36) then becomes

$$P_{\text{ideal}}(\phi) = P_{\text{ideal}}(\theta_A, \theta_B) = \frac{1}{2}(\cos\theta_A\cos\theta_B + \sin\theta_A\sin\theta_B)^2 = \frac{1}{2}\cos^2(\theta_A - \theta_B) = \frac{1}{2}\cos^2\phi. \tag{6.39}$$

If ε_a and ε_b represent the respective transmittances for light polarized along the polarizer axes, then the predicted probability of equation (6.39) changes to

$$P_{\text{actual}}(\phi) = \frac{N(\phi)}{N_0} = \frac{1}{2}\varepsilon_a\varepsilon_b\cos^2\phi. \tag{6.40}$$

Hence, we can use equation (6.40) to compute a quantum mechanical prediction for δ as follows [23]:

$$\delta = \left| \frac{N(22.5°) - N(67.5°)}{N_0} \right| - \frac{1}{4} = \frac{\varepsilon_a\varepsilon_b}{2\sqrt{2}} - \frac{1}{4}. \tag{6.41}$$

If we plug $\delta = 0$ in equation (6.41), we get $\varepsilon_a\varepsilon_b = 0.71$ or $\sqrt{\varepsilon_a\varepsilon_b} = 0.84$. This implies that the polarizer transmittances should have a geometric mean of at least 0.84 to demonstrate a violation of Freedman's inequality. Such a requirement on the experimental apparatus is unique to Freedman's test. Additionally, finding polarizers with such high transmittances is difficult. Luckily for us, we are not using polarizers. Instead, we use a combination of an HWP oriented at half the analysis angle, followed by a PBS. Observing one of the faces of the PBS in this setup is operationally equivalent to using a polarizer. In other words, if we only turn on detectors A and B in our setup (see figure 6.3), the HWPs followed by PBSs will act as polarizers. Using the assembly of HWP followed by a PBS gives higher transmittance than polarizers. Furthermore, cleaning the optical components may also help in recovering high transmittances.

6.4.3 The experiment

We begin by generating Bell state $|\Phi^+\rangle$ by adjusting the pump beam HWP and the quartz plate as discussed in section 6.3:

$$|\Phi^+\rangle = \frac{1}{\sqrt{2}}(|IIII\rangle + |VV\rangle). \tag{6.42}$$

The coincidence counts[2] $N(22.5°)$, $N(67.5°)$ and N_0 are measured over equally timed intervals. To determine N_0, the coincidence counts N_A and N_B are measured with one set of HWP and PBS removed at a time and the other oriented to achieve maximum counts. The transmittances are calculated as $\varepsilon_A = 2N_A/N_0$ and $\varepsilon_B = 2N_B/N_0$ [23]. There is a factor of 2 in these expressions because for the Bell state $|\Phi^+\rangle$, analyzing the downconverted beams reduces the coincidence counts to roughly half the original value N_0. The reduced counts N_A and N_B are proportional to the respective transmittances.

Our sets of HWP followed by the PBS gives us transmittances $\varepsilon_A = 0.975$ and $\varepsilon_B = 0.995$. With equation (6.41), the quantum mechanical prediction for δ is calculated as

---

[2] Note that we are writing the analysis angles here. The HWPs are to be oriented at half these angles.

Table 6.2. Results of Freedman's test of locality. R represents the rate (per second) of photocounts. Multiplying R with the integration time gives the total number of counts, N.

Integration time	$R(22.5°)$	$R(67.5°)$	R_0	δ
30 s	2987 ± 8	522 ± 4	7281 ± 16	$0.088\,64 \pm 0.0004$

$$\delta = \frac{\varepsilon_a \varepsilon_b}{2\sqrt{2}} - \frac{1}{4} = \frac{0.975 \times 0.995}{2\sqrt{2}} - \frac{1}{4} = 0.0932. \tag{6.43}$$

This predicts that with our apparatus, we should be able to violate Freedman's inequality by a significant margin.

The experimental results clearly show a violation of Freedman's inequality and rule out local hidden variable models. In other words, entangled systems demonstrate nonlocality. We recorded data for 30 s, and the results violate Freedman's inequality with a confidence interval of 219 standard error deviations. The results are summarized in table 6.2.

We have seen that we need to use just two single-photon detectors and perform only three measurements to perform a test of Freedman's inequality. This is perhaps the least complicated experiment for testing locality in an entanglement system. However, this test does have certain limitations. For instance, it uses the no-enhancement and the rotational invariance assumptions (which are related to the properties of the analyzers) while deriving the inequality. A stronger test of nonlocality, which we will discuss in the next section, will not require these additional assumptions. The additional assumptions are said to create 'loopholes' that make Freedman's test a rather weak test of locality. Nonetheless, the minimalist hardware requirements of this nonlocality test still make it an excellent experiment for teaching purposes.

6.5 NL2: CHSH test of locality

Experimental violation of CHSH inequality is a significant test of locality and is more laborious regarding the number of measurements. This experiment could be performed with two detectors, but that would require measurements over a larger set of apparatus settings [33]. Our four-detector setup reduces the number of trials for one complete experiment, which is why we prefer it. Let us first look at the inequality from a purely formal and conceptual perspective.

6.5.1 The CHSH inequality

We follow the approach presented in reference [16] to derive the CHSH inequality. Consider, once again, the Alice–Bob experiment of figure 6.1. For a hidden variable theory with a hidden variable denoted as λ, we can posit the probability distribution

$$p(\lambda) \geqslant 0, \tag{6.44}$$

which, of course, is normalized,

$$\int p(\lambda)d\lambda = 1. \tag{6.45}$$

Locality assumption is included in the argument as follows. For a photon going to Alice, the measurement outcome is completely determined by $A(\alpha, \lambda)$, a function of the polarization angle α and the hidden variable λ. Its value is taken to be $+1$ if the photon is polarized along the analysis angle $\alpha = \theta_A$ and -1 if it is polarized orthogonal to it $\alpha = \theta_{A'}$. These conventions for θ_A and $\theta_{A'}$ have been defined in detail in section 6.3.2. Similarly, Bob's photon measurement can be described by $B(\beta, \lambda)$, which takes the value of $+1$ for $\beta = \theta_B$ and -1 for $\beta = \theta_{B'}$. A hidden variable theory with hidden variable λ therefore specifies $p(\lambda)$, $A(\theta_A, \lambda)$ and $B(\theta_B, \lambda)$.

We know that the probability of a specific outcome averaged over an ensemble of photon pairs is given by an integral. Therefore, the probabilities of finding photon pairs for different polarization combinations are given by the following expressions:

$$P(\theta_A, \theta_B) = \int \left(\frac{1 + A(\alpha, \lambda)}{2}\right)\left(\frac{1 + B(\beta, \lambda)}{2}\right)p(\lambda)d\lambda, \tag{6.46}$$

$$P(\theta_A, \theta_{B'}) = \int \left(\frac{1 + A(\alpha, \lambda)}{2}\right)\left(\frac{1 - B(\beta, \lambda)}{2}\right)p(\lambda)d\lambda, \tag{6.47}$$

$$P(\theta_{A'}, \theta_B) = \int \left(\frac{1 - A(\alpha, \lambda)}{2}\right)\left(\frac{1 + B(\beta, \lambda)}{2}\right)p(\lambda)d\lambda, \tag{6.48}$$

$$P(\theta_{A'}, \theta_{B'}) = \int \left(\frac{1 - A(\alpha, \lambda)}{2}\right)\left(\frac{1 - B(\beta, \lambda)}{2}\right)p(\lambda)d\lambda. \tag{6.49}$$

The terms of the form $(1 \pm A)/2$ and $(1 \pm B)/2$ modify the overall contribution of the terms $A(\alpha, \lambda)$ and $B(\beta, \lambda)$ from the intervals $\{-1, 1\}$ to $\{0, 1\}$. Moreover, they will prove useful for obtaining a convenient expression for a function denoted as $E(\alpha, \beta)$.

To make intuitive sense of the expressions in equations (6.46)–(6.49), consider the integral for $P(\theta_A, \theta_B)$ which determines the probability of detecting Alice's and Bob's photons polarized along θ_A and θ_B, respectively. For a specific value of λ, Alice's term $(1 + A(\alpha, \lambda))/2$ will be equal to 1 or 0 corresponding to whether her photon is polarized along θ_A or $\theta_{A'}$. A similar explanation goes for Bob's term $(1 + B(\beta, \lambda))/2$. Both terms will only contribute to the overall probability when Alice's photon is polarized along θ_A and Bob's photon is polarized along θ_B. Hence, the integral calculates $P(\theta_A, \theta_B)$. We can look at the other integrals likewise.

Using equations (6.46)–(6.49), we can show that

$$E(\theta_A, \theta_B) \equiv P(\theta_A, \theta_B) + P(\theta_{A'}, \theta_{B'}) - P(\theta_A, \theta_{B'}) - P(\theta_{A'}, \theta_B) = \int A(\alpha, \lambda)B(\beta, \lambda)p(\lambda)d\lambda. \tag{6.50}$$

The variable $E(\theta_A, \theta_B)$ is, in fact, the expected outcome of local measurements that determine Alice's photon to be polarized along θ_A and Bob's along θ_B. Let us also define s that tells us about the polarization correlation in one pair of photons in terms of the four angles θ_{A1}, θ_{A2}, θ_{B1} and θ_{B2} [16]:

$$
\begin{aligned}
s &= A(\theta_{A1}, \lambda)B(\theta_{B1}, \lambda) - A(\theta_{A1}, \lambda)B(\theta_{B2}, \lambda) + A(\theta_{A2}, \lambda)B(\theta_{B1}, \lambda) + A(\theta_{A2}, \lambda,)B(\theta_{B2}, \lambda) \\
&= A(\theta_{A1}, \lambda)(B(\theta_{B1}, \lambda) - B(\theta_{B2}, \lambda)) + A(\theta_{A2}, \lambda)(B(\theta_{B1}, \lambda,) + B(\theta_{B2}, \lambda,)).
\end{aligned}
\tag{6.51}
$$

It is not hard to see that s can only take the values $+2$ or -2. If we have an ensemble of photons, the average of s can be calculated as

$$
\begin{aligned}
S(\theta_{A1}, \theta_{A2}, \theta_{B1}, \theta_{B2}) &\equiv \langle s|s \rangle = \int s(\lambda, \theta_{A1}, \theta_{A2}, \theta_{B1}, \theta_{B2})p(\lambda)\mathrm{d}(\lambda) \\
&= E(\theta_{A1}, \theta_{B1}) - E(\theta_{A1}, \theta_{B2}) + E(\theta_{B2}, \theta_{B1}) + E(\theta_{A2}, \theta_{B2}).
\end{aligned}
\tag{6.52}
$$

Since s can have only the values ± 2, S, which is its average, must satisfy

$$
|S| \leqslant 2.
\tag{6.53}
$$

This inequality is called the CHSH inequality, and any local hidden variable theory must obey it. Next, we look at the quantum mechanical prediction.

6.5.2 Quantum mechanical prediction for the CHSH test

Consider the bipartite Bell state (6.35) reproduced here:

$$
|\Phi^+\rangle = \frac{1}{\sqrt{2}}(|HH\rangle + |VV\rangle).
\tag{6.54}
$$

We revisit the rotated $\{|H\rangle, |V\rangle\}$ basis that we alluded to in the previous section. Consider Alice's setup; see figure 6.3. If we measure polarization in the $\{|H\rangle, |V\rangle\}$ basis that is rotated anticlockwise at an angle θ_A with respect to the horizontal, then the basis states are given by

$$
|\theta_A\rangle = -\sin\theta_A|H\rangle + \cos\theta_A|V\rangle,
\tag{6.55}
$$

$$
|\theta_{A'}\rangle = \cos\theta_A|H\rangle + \sin\theta_A|V\rangle.
\tag{6.56}
$$

Similarly, for Bob with a basis rotated at θ_B with respect to the horizontal, the basis states are given by

$$
|\theta_B\rangle = -\sin\theta_B|H\rangle + \cos\theta_B|V\rangle,
\tag{6.57}
$$

$$
|\theta_{B'}\rangle = \cos\theta_B|H\rangle + \sin\theta_B|V\rangle.
\tag{6.58}
$$

The measurement probabilities are then given by

$$
P(\theta_A, \theta_B) = |(\langle\theta_A|\langle\theta_B|)|\Phi^+\rangle|^2 = \frac{1}{2}\cos^2(\theta_A - \theta_B).
\tag{6.59}
$$

Similarly, we have

$$
P(\theta_{A'}, \theta_{B'}) = \frac{1}{2}\cos^2(\theta_A - \theta_B) \quad \text{and}
\tag{6.60}
$$

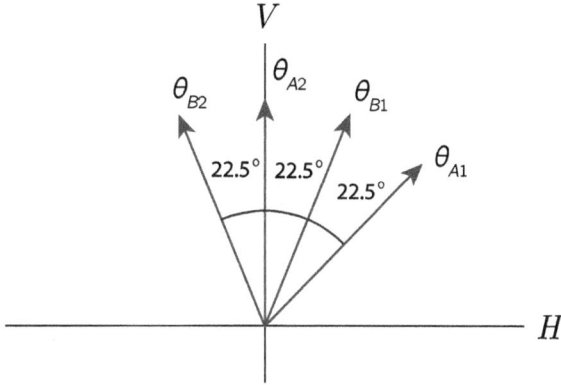

Figure 6.5. Relations between analysis angles for achieving maximal violation of CHSH inequality. According to equation (6.63), the absolute values of the angles do not matter—only the relative orientations matter.

$$P(\theta_{A'}, \theta_B) = P(\theta_A, \theta_{B'}) = \frac{1}{2}\sin^2(\theta_A - \theta_B). \tag{6.61}$$

Plugging equations (6.59)–(6.61) into equation (6.50), we obtain

$$E(\theta_A, \theta_B) = P(\theta_A, \theta_B) + P(\theta_{A'}, \theta_{B'}) - P(\theta_A, \theta_{B'}) - P(\theta_{A'}, \theta_B) = \cos(2(\theta_A - \theta_B)). \tag{6.62}$$

Using the above equation with equation (6.52), we ultimately get the following expression for S:

$$\begin{aligned} S &= E(\theta_{A1}, \theta_{B1}) - E(\theta_{A1}, \theta_{B2}) + E(\theta_{A2}, \theta_{B1}) + E(\theta_{A2}, \theta_{B2}) \\ &= \cos(2(\theta_{A1} - \theta_{B1})) - \cos(2(\theta_{A1} - \theta_{B2})) \\ &\quad + \cos(2(\theta_{A2} - \theta_{B1})) + \cos(2(\theta_{A2} - \theta_{B2})). \end{aligned} \tag{6.63}$$

It can be shown that for angles $\theta_{A1} = -45°$, $\theta_{A2} = 0°$, $\theta_{B1} = -22.5°$ and $\theta_{B2} = 22.5°$, we get $S = 2\sqrt{2}$. This numerical figure provides the maximal possible violation of CHSH inequality. Other analysis angles may be chosen instead, but it must be noted that we will get the maximal violation for the angles whose relative disposition [16] is as stated and is shown in figure 6.5.

It is important to mention that $S = 2\sqrt{2}$ holds only for the Bell state given by equation (6.54). Other states produce lower values of S. Experimentally, it is very hard to perfectly produce the state represented by equation (6.54). Therefore, we expect to obtain nonideal results. Nevertheless, if we can experimentally show that $S > 2$, then it will be a conclusive violation of Bell's inequality, negating all local hidden variable theories.

6.5.3 The experiment

We generate the Bell state following the procedure outlined in section 6.3.3. For this Bell state, the analysis angles are chosen to be $\theta_{A1} = -45°$, $\theta_{A2} = 0°$, $\theta_{B1} = -22.5°$, and $\theta_{B2} = 22.5°$.

Table 6.3. Values of E obtained using equation (6.62). These values are used to find the S value in table 6.4.

Integration time (s)	$E(\theta_{A1}, \theta_{B1})$	$E(\theta_{A1}, \theta_{B2})$	$E(\theta_{A2}, \theta_{B1})$	$E(\theta_{A2}, \theta_{B2})$
30	0.593 ± 0.003	-0.538 ± 0.003	0.722 ± 0.002	0.755 ± 0.002
60	0.593 ± 0.002	-0.538 ± 0.002	0.722 ± 0.002	0.756 ± 0.002
120	0.591 ± 0.001	-0.536 ± 0.001	0.709 ± 0.001	0.768 ± 0.001

Table 6.4. Results of the CHSH test of locality. The column labeled 'Confidence' lists the number of standard deviations σ by which the experimental results violate $|S| \leqslant 2$. The violation from the localistic theories becomes stronger for longer integration times.

Integration time (s)	S	Confidence
30	2.608 ± 0.005	118σ
60	2.610 ± 0.004	158σ
120	2.604 ± 0.003	219σ

Since we are using the four-detector scheme with HWPs and beam splitters instead of polarizers, we orient the detector HWPs to half the analysis angles and then detect coincidence counts for a certain integration time (table 6.3.). We record three datasets and summarize the experimental results in table 6.4. These results clearly show that the system under study violates locality, agreeing with the quantum mechanical prediction ($S > 2$).

The last test for entanglement that we will discuss is the Hardy's test. Hardy's experiment is designed to be tested with a different entangled state from the usual Bell state.

6.6 NL3: Hardy's test of locality

We shall observe that the implementation of Hardy's test [24] is quite similar to that of the CHSH test of locality [16, 29]. The apparatus used in both experiments is the same, but there are some differences in how these experiments are executed. For instance, different entangled states are generated with the downconverted photons, and the analysis angles are different. In Hardy's test, we are interested in analyzing the coincidence counts to determine a quantity that we shall represent as H.

One may question the preference of performing one test to the other. Even though it is incrementally more involved than Freedman's test, Hardy's test is considerably easier to comprehend than the CHSH test. Let us first derive the quantity H that we shall use as our measure to determine violation of locality.

6.6.1 The Hardy inequality

We derive the Hardy inequality following the approach detailed in reference [30]. For this derivation, let us go back to Alice and Bob (figure 6.1), who are interested in the joint probability $P(\theta_A, \theta_B)$ of the event when Alice detects a photon polarized along the angle θ_A, and Bob detects a photon polarized along the angle θ_B.

According to quantum mechanics, the aforementioned probability is encoded in the composite wavefunction $|\psi\rangle$. On the other hand, in localistic theories, we propose the presence of a hidden variable λ, which determines the joint probabilities. As mentioned earlier in this chapter, this is a variable that cannot be measured, yet it encodes the probabilities that we want to determine. In a hidden variable theory,

$$P(\theta_A, \theta_B) \equiv \sum_\lambda P(\theta_A, \theta_B, \lambda) = \sum_\lambda P(\theta_A, \theta_B|\lambda)P(\lambda) \tag{6.64}$$

because there could be multiple variables specifying the joint probability. This is the reality assumption. The probability can also be written as

$$P(\theta_A, \theta_B) = \sum_\lambda P(\theta_A, \theta_B, \lambda) = \sum_\lambda P(\theta_A|\theta_B, \lambda)P(\theta_B, \lambda)$$
$$= \sum_\lambda P(\theta_A|\theta_B, \lambda)P(\theta_B|\lambda)P(\lambda). \tag{6.65}$$

Another assumption here is that the probability distributions we are dealing with are all classical probability distributions, i.e., the probabilities are real, nonnegative, and normalized.

Furthermore, the locality assumption tells us that the measurement made by Bob cannot affect the measurement made by Alice and vice versa. Mathematically, this can be stated as

$$P(\theta_A|\theta_B, \lambda) = P(\theta_A|\lambda) \text{ and} \tag{6.66}$$

$$P(\theta_B|\theta_A, \lambda) = P(\theta_B|\lambda). \tag{6.67}$$

Therefore, assuming locality, the joint probability of equation (6.65) becomes

$$P(\theta_A, \theta_B) = \sum_\lambda P(\theta_A|\lambda)P(\theta_B|\lambda)P(\lambda). \tag{6.68}$$

Normalization of probability distributions tells us that

$$P(\theta_A|\lambda) + P(\theta_{A'}|\lambda) = 1, \tag{6.69}$$

meaning that if the source produces photons specified by the hidden variable λ, then two possibilities exist for Alice's photon: the photon is polarized either along θ_A or along $\theta_{A'}$ (i.e., perpendicular to θ_A). The original joint probability can then be treated as follows:

$$P(\theta_{A1}, \theta_{B1}) = \sum_{\lambda} P(\theta_{A1}|\lambda)P(\theta_{B1}|\lambda)P(\lambda)$$

$$= \sum_{\lambda} P(\theta_{A1}|\lambda)P(\theta_{B1}|\lambda)(P(\theta_{A2}|\lambda) + P(\theta_{A2'}|\lambda))P(\lambda)$$

$$= \sum_{\lambda} P(\theta_{A1}|\lambda)P(\theta_{B1}|\lambda)P(\theta_{A2}|\lambda)P(\lambda)$$

$$+ \sum_{\lambda} P(\theta_{A1}|\lambda)P(\theta_{B1}|\lambda)P(\theta_{A2'}|\lambda)P(\lambda)$$

$$= \sum_{\lambda} P(\theta_{A1}|\lambda)P(\theta_{B1}|\lambda)P(\theta_{A2}|\lambda)(P(\theta_{B2}|\lambda) + P(\theta_{B2'}|\lambda))P(\lambda) \qquad (6.70)$$

$$+ \sum_{\lambda} P(\theta_{A1}|\lambda)P(\theta_{B1}|\lambda)P(\theta_{A2'}|\lambda)P(\lambda)$$

$$= \sum_{\lambda} P(\theta_{A1}|\lambda)P(\theta_{B1}|\lambda)P(\theta_{A2}|\lambda)P(\theta_{B2}|\lambda)P(\lambda)$$

$$+ \sum_{\lambda} P(\theta_{A1}|\lambda)P(\theta_{B1}|\lambda)P(\theta_{A2}|\lambda)P(\theta_{B2'}|\lambda)P(\lambda)$$

$$+ \sum_{\lambda} P(\theta_{A1}|\lambda)P(\theta_{B1}|\lambda)P(\theta_{A2'}|\lambda)P(\lambda).$$

Keeping in mind the basic axioms that all probabilities need to be real, positive, and less than 1, the three terms in equation (6.70) obey the inequalities

$$\sum_{\lambda} P(\theta_{A1}|\lambda)P(\theta_{B1}|\lambda)P(\theta_{A2}|\lambda)P(\theta_{B2}|\lambda)P(\lambda) \leqslant \sum_{\lambda} P(\theta_{A2}|\lambda)P(\theta_{B2}|\lambda)P(\lambda)$$
$$= P(\theta_{A2}, \theta_{B2}), \qquad (6.71)$$

$$\sum_{\lambda} P(\theta_{A1}|\lambda)P(\theta_{B1}|\lambda)P(\theta_{A2}|\lambda)P(\theta_{B2'}|\lambda)P(\lambda) \leqslant \sum_{\lambda} P(\theta_{A2}|\lambda)P(\theta_{B2'}|\lambda)P(\lambda)$$
$$= P(\theta_{A2}, \theta_{B2'}), \qquad (6.72)$$

$$\sum_{\lambda} P(\theta_{A1}|\lambda)P(\theta_{B1}|\lambda)P(\theta_{A2'}|\lambda)P(\lambda) \leqslant \sum_{\lambda} P(\theta_{B1}|\lambda)P(\theta_{A2'}|\lambda)P(\lambda)$$
$$= P(\theta_{A2'}, \theta_{B1}). \qquad (6.73)$$

Putting these inequalities into equation (6.70) yields the inequality

$$P(\theta_{A1}, \theta_{B1}) \leqslant P(\theta_{A2}, \theta_{B2}) + P(\theta_{A1}, \theta_{B2'}) + P(\theta_{A2'}, \theta_{B1}). \qquad (6.74)$$

This is called the Bell-Clauser–Horne inequality [32], and any localistic theory should obey it. If this inequality is violated, quantum mechanical predictions jeopardizing a localistic theory are validated, and nature is shown to violate locality at the fundamental level.

The above inequality can also be written as

$$P(\theta_{A2}, \theta_{B2}) \geqslant P(\theta_{A1}, \theta_{B1}) - P(\theta_{A1}, \theta_{B2'}) - P(\theta_{A2'}, \theta_{B1}). \qquad (6.75)$$

This involves four independent angles $(\theta_{A1}, \theta_{A2}, \theta_{B1}, \theta_{B2})$. If we choose these angles to be $\theta_{A1} = \beta$, $\theta_{B1} = -\beta$, $\theta_{A2} = -\alpha$, and $\theta_{B2} = \alpha$, then equation (6.75) becomes

$$P(-\alpha, \alpha) \geqslant P(\beta, -\beta) - P(\beta, \alpha') - P(-\alpha', -\beta). \tag{6.76}$$

We define a quantity H as

$$H \equiv P(\beta, -\beta) - P(\beta, \alpha') - P(-\alpha', -\beta) - P(-\alpha, \alpha). \tag{6.77}$$

Then, if we experimentally obtain $H \leqslant 0$, then locality is satisfied. On the contrary, it will be a violation of locality if we obtain $H > 0$. Let us see how a strictly quantum mechanical theory allows H to be positive, violating locality.

6.6.2 Quantum mechanical prediction for Hardy's test

Consider a bipartite system defined by the general state

$$|\psi\rangle = A|HH\rangle + B|VV\rangle, \tag{6.78}$$

with A and B being real numbers and the state being normalized (i.e., $|A|^2 + |B|^2 = 1$).

If the source generates photons in the state $|\psi\rangle$, then the joint probability of the event that Alice detects a photon polarized along the angle θ_{Ai} and Bob detects a photon polarized along the angle θ_{Bj} is given by the Born rule:

$$P(\theta_{Ai}, \theta_{Bj}) = |(\langle\theta_{Ai}|\langle\theta_{Bj}|)|\psi\rangle|^2. \tag{6.79}$$

The measurement basis is chosen by orienting a HWP at the appropriate angle in the path of each beam. If Alice's and Bob's analysis angles are θ_{Ai} and θ_{Bj} with respect to the horizontal, then we can use the formulation

$$|\theta\rangle = \cos\theta|H\rangle + \sin\theta|V\rangle \tag{6.80}$$

to determine the joint probability as

$$\begin{aligned} P(\theta_{Ai}, \theta_{Bj}) &= |(\langle H|\cos\theta_{Ai} + \langle V|\sin\theta_{Ai})(\langle H|\cos\theta_{Bj} + \langle V|\sin\theta_{Bj}) \\ &\quad (A|HH\rangle + B|VV\rangle)|^2 \\ &= (A\cos\theta_{Ai}\cos\theta_{Bj} + B\sin\theta_{Ai}\sin\theta_{Bj})^2. \end{aligned} \tag{6.81}$$

Using equation (6.81), one can verify that maximal violation of Hardy's inequality can be achieved if either of the following two states is generated to perform the experiment [24, 30]:

$$|\psi_1\rangle = \sqrt{0.8}|HH\rangle + \sqrt{0.2}|VV\rangle, \tag{6.82}$$

$$|\psi_2\rangle = \sqrt{0.2}|HH\rangle + \sqrt{0.8}|VV\rangle. \tag{6.83}$$

For $|\psi_1\rangle$, the analysis angles are $\alpha = 55°$ and $\beta = 71°$, and for $|\psi_2\rangle$, the analysis angles are $\alpha = 35°$ and $\beta = 19°$ [24]. We can check that for $A = \sqrt{0.2}$ and $B = \sqrt{0.8}$ in equation (6.78), choosing $\alpha = 35°$ and $\beta = 19°$, equations (6.82) and (6.83) are called Hardy states, and result in $H = 0.093 > 0$. Hence, mathematical analysis shows that the system will violate the assumption of locality. Let us begin the experimental discussion by first going through the procedure of generating the Hardy states.

6.6.3 Tuning the Hardy state

For this experiment, we shall follow the method outlined in reference [24] and generate the entangled state $|\psi_2\rangle$ given in equation (6.83). First, we orient both the detector HWPs at $0°$. In this setting, on the one hand N_{AB} coincidence counts correspond to detections of $|HH\rangle$ photons, and on the other hand $N_{A'B'}$ coincidences correspond to the detection of $|VV\rangle$ photons. Then, we modify the orientation of the pump beam HWP until the ratio of the aforementioned coincidence counts is approximately 1:4. We then set the detector HWPs in Alice's and Bob's paths to monitor $N_{AB}(-\alpha, \alpha)$ and carefully tweak the quartz plate tilt to minimize these coincidences. This minimization corresponds to minimizing φ in state (6.1). This can be tested by minimizing the overlap between $|-\alpha\rangle \otimes |\alpha\rangle$ and $\sqrt{0.2}\,|HH\rangle + \sqrt{0.8}\,e^{i\phi}|VV\rangle$ and verifying that the overlap is indeed minimal when $\phi = 0$. The overlap is proportional to the count rate $N_{AB}(-\alpha, \alpha)$.

Once this much has been done, the generated state should be fairly well-tuned according to the requirement. A final check would be to adjust the measurement setup to see that the probabilities $P(\beta, \alpha')$ and $P(-\alpha', -\beta)$ are both minimum. As detailed in the previous paragraph, the state can be further fine-tuned by adjusting the pump beam HWP and the quartz plate. Essentially, the goal is to get minimum possible values of $P(-\alpha, \alpha)$, $P(\beta, \alpha')$ and $P(-\alpha', -\beta)$, and hence get an increased value of H (see equation (6.77)).

6.6.4 The experiment

Now that we have tuned our setup to generate one of the two Hardy states, we are ready to perform the test for entanglement. For the state given by equation (6.83), the analysis angles are $\alpha = 35°$, $\beta = 19°$. Since our analyzers comprise HWPs and beam splitters instead of polarizers, the HWPs need to be rotated at half the analysis angles with respect to the horizontal to produce the desired functionality. The coincidence counts were recorded for an integration time of two minutes for each pair of analysis angles. The required probabilities came out to be

$$P(\beta, -\beta) = 0.1122 \pm 0.0004, \tag{6.84}$$

$$P(\beta, \alpha') = 0.0217 \pm 0.0002, \tag{6.85}$$

$$P(-\alpha', -\beta) = 0.0187 \pm 0.0002, \quad \text{and} \tag{6.86}$$

$$P(-\alpha, \alpha) = 0.0329 \pm 0.0002, \tag{6.87}$$

ultimately leading to the value $H = 0.0389 \pm 0.0005$. The experimental result violates the inequality $H \leqslant 0$ by 71 standard deviations. This is a conclusive violation of locality.

6.7 Conclusion

Arguably, one of the most attractive aspects of physics is pondering over the nature of physical reality. Many Bell tests have been performed and re-performed to this day, each version seemingly better than the previous ones and closing more loopholes. While there is frontier research in advanced ideas such as entanglement in quantum networks [34], measuring entanglement using one-photon measurement [35], and entanglement of fundamentally different kinds of particles [36], here we have discussed three very accessible versions of the Bell test—experiments that are a touchstone of some of the deepest investigations of the ultimate nature of this world's (quantum) reality. Entanglement is a captivating phenomenon, and the applications it promises, such as quantum secure communication and quantum teleportation, make it even more fascinating. In the next chapter, experiment Q+NL will demonstrate nonlocal effects that combine entanglement with quantum interference.

References

[1] Schrödinger E 1935 Discussion of probability relations between separated systems *Mathematical Proc. of the Cambridge Philosophical Society* **vol 31** (Cambridge: Cambridge University Press) pp 555–63

[2] Einstein A, Podolsky B and Rosen N 1935 Can quantum-mechanical description of physical reality be considered complete? *Phys. Rev.* **47** 777

[3] Ekert A K 1991 Quantum cryptography based on Bell's theorem *Phys. Rev. Lett.* **67** 661

[4] Ekert A K, Rarity J G, Tapster P R and Palma G M 1992 Practical quantum cryptography based on two-photon interferometry *Phys. Rev. Lett.* **69** 1293

[5] Bennett C H, Brassard G, Crépeau C, Jozsa R, Peres A and Wootters W K 1993 Teleporting an unknown quantum state via dual classical and Einstein-Podolsky-Rosen channels *Phys. Rev. Lett.* **70** 1895

[6] Bennett C H and Wiesner S J 1992 Communication via one- and two-particle operators on Einstein-Podolsky-Rosen states *Phys. Rev. Lett.* **69** 2881

[7] Hidary J D 2019 *Quantum Computing: An Applied Approach* (Berlin: Springer)

[8] Stapp H P 1972 The Copenhagen interpretation *Am. J. Phys.* **40** 1098–116

[9] Bell J S 1964 On the Einstein-Podolsky-Rosen paradox *Phys. P. Fiz.* **1** 195

[10] Bell J S and Bell J S 2004 *Speakable and Unspeakable in Quantum Mechanics: Collected Papers on Quantum Philosophy* (Cambridge: Cambridge University Press)

[11] Clauser J F, Horne M A, Shimony A and Holt R A 1969 Proposed experiment to test local hidden-variable theories *Phys. Rev. Lett.* **23** 880

[12] Hardy L 1993 Nonlocality for two particles without inequalities for almost all entangled states *Phys. Rev. Lett.* **71** 1665

[13] Freedman S J and Clauser J F 1972 Experimental test of local hidden-variable theories *Phys. Rev. Lett.* **28** 938

[14] Pan J-W, Bouwmeester D, Daniell M, Weinfurter H and Zeilinger A 2000 Experimental test of quantum nonlocality in three-photon Greenberger-Horne-Zeilinger entanglement *Nature* **403** 515

[15] Torgerson J R, Branning D, Monken C H and Mandel L 1995 Experimental demonstration of the violation of local realism without Bell inequalities *Phys. Lett.* A **204** 323–8

[16] Dehlinger D and Mitchell M W 2002 Entangled photons, nonlocality, and Bell inequalities in the undergraduate laboratory *Am. J. Phys.* **70** 903–10

[17] Galvez E J 2014 Resource letter SPE-1: single-photon experiments in the undergraduate laboratory *Am. J. Phys.* **82** 1018–28

[18] Greenberger D M, Horne M A and Zeilinger A 1989 Going beyond Bell's theorem Bell's Theorem *Quantum Theory and Conceptions of the Universe* (Berlin: Springer) pp 69–72

[19] Mermin N D 1990 Quantum mysteries revisited *Am. J. Phys.* **58** 731–4

[20] Mermin N D 1995 The best version of Bell's theorem a *Ann. N. Y. Acad. Sci.* **755** 616–23

[21] Di Giuseppe G, De Martini F and Boschi D 1997 Experimental test of the violation of local realism in quantum mechanics without Bell inequalities *Phys. Rev.* A **56** 176

[22] White A G, James D F V, Eberhard P H and Kwiat P G 1999 Nonmaximally entangled states: production, characterization, and utilization *Phys. Rev. Lett.* **83** 3103

[23] Brody J and Selton C 2018 Quantum entanglement with Freedman's inequality *Am. J. Phys.* **86** 412–6

[24] Carlson J A, Olmstead M D and Beck M 2006 Quantum mysteries tested: an experiment implementing Hardy's test of local realism *Am. J. Phys.* **74** 180–6

[25] Mermin N D 1994 Quantum mysteries refined *Am. J. Phys.* **62** 880–7

[26] Kwiat P G and Hardy L 2000 The mystery of the quantum cakes *Am. J. Phys.* **68** 33–6

[27] Branning D 1997 Does nature violate local realism? A theoretical battle–asking whether quantum mechanics provides a complete description of nature–comes under experimental scrutiny *Am. Sci.* **85** 160–7

[28] Kwiat P G, Waks E, White A G, Appelbaum I and Eberhard P H 1999 Ultrabright source of polarization-entangled photons *Phys. Rev.* A **60** R773

[29] Dehlinger D and Mitchell M W 2002 Entangled photon apparatus for the undergraduate laboratory *Am. J. Phys.* **70** 898–902

[30] Beck M 2012 *Quantum Mechanics: Theory and Experiment* (Oxford: Oxford University Press)

[31] Aspect A, Grangier P and Roger G 1981 Experimental tests of realistic local theories via Bell's theorem *Phys. Rev. Lett.* **47** 460

[32] Clauser J F and Horne M A 1974 Experimental consequences of objective local theories *Phys. Rev.* D **10** 526

[33] Negre A, Mathevet R, Chalopin B and Massenot S 2023 Unexpected optimal measurement protocols in Bell's inequality violation experiments *Am. J. Phys.* **91** 64–73

[34] Abiuso P, Kriváchy T, Boghiu E-C, Renou M-O, Pozas-Kerstjens A and Acín A 2022 Single-photon nonlocality in quantum networks *Phys. Rev. Res.* **4** L012041

[35] Lemos G B, Lapkiewicz R, Hochrainer A, Lahiri M and Zeilinger A 2023 One-photon measurement of two-photon entanglement *Phys. Rev. Lett.* **130** 090202

[36] Hannegan J, Siverns J D and Quraishi Q 2022 Entanglement between a trapped-ion qubit and a 780-nm photon via quantum frequency conversion *Phys. Rev.* A **106** 042441

IOP Publishing

Quantum Mechanics in the Single-Photon Laboratory (Second Edition)

Muhammad Sabieh Anwar, Faizan-e-Ilahi, Syed Bilal Hyder and Muhammad Hamza Waseem

Chapter 7

Quantum interference and quantum erasure

This book has discussed various facets of single photons and their quantum nature. In chapter 4, we generated a beam of single photons and utilized the second-order decoherence $g^{(2)}(0)$ to show that the photon view eludes a classical description of light. A photon could only be detected once. This is commensurate with our meaning of the word 'particle'. The experiments in chapter 5 equipped us with the ability to manipulate and measure the polarization of single photons. Moreover, chapter 6 described experiments that credibly demonstrated violation of Bell's inequalities, affirming that entangled single photons defy locality.

Furthering these ideas, this chapter describes two closely related experiments which illustrate wave-particle duality, i.e., grainy photons can also interfere with themselves, even though they are coming into an interferometer one at a time. This interference can be suppressed or reclaimed even after the photon has trespassed the interferometer. Finally, the choice for interference/no-interference can also be made nonlocally. This is too much in a single paragraph. We will now unpack these investigations. The experiments are given below:

- Q5: Single-photon interference and quantum erasure
- Q+NL: Nonlocal quantum erasure

7.1 Q5: Single-photon interference and quantum erasure

In experiment Q5, we will make the single photons traverse through an interferometer, allowing us to observe the interference of single photons. In fact, some laboratories have simultaneously implemented both second-order correlation measurements (which is our experiment Q2) and interferometric measurements (experiment Q5) within the same experiment [1, 2]. However, for the sake of simplicity, we have separated the two and will just focus on the interference part in this experiment. Before providing details here, let us have a general overview of interference and quantum erasure, as well as a brief survey of related experiments. The classical version of this experiment was a subject of section 2.6.

If light is made to pass through an interferometer, then the visibility[1] is dependent on the extent to which the 'which-way' or 'which-path' information is available to the observer. While delineating Experiment C3, we found that, in principle, if the experimenter can determine the path that light traverses through in the interferometer, then no interference will be observed. On the other hand, if the experimenter does not know about the path of light (and cannot even, in principle, obtain which-way information), interference fringes with very high visibility will be observed. In the more general situation, if partial path information is available, partial interference can be observed [3]. We can investigate which-path information using an interferometer for various practical implementations, including Young's original double-slit experiment [4].

Some experiments have been proposed in which the experimenter can switch between complete information and no information about the path of light by making minor changes in the experimental setup [1, 3, 5]. In such experiments, if the which-path information is rendered unavailable, then it is said to be *erased*. The erasure results in the reinstatement of interference fringes of high visibility. Such an interferometer is usually termed as 'quantum eraser'. In a highly counter-intuitive twist to reality, the erasure of which-path information can be materialized even *after* the photon has already passed through the interferometer. This is like altering the outcome of an experiment after it has occurred.

There are a number of ways to make a quantum eraser. For instance, as discussed in chapter 2 (Experiment C3), we can insert polarizers into the two paths of a Mach–Zehnder interferometer, tagging the paths of the corresponding beams through their definite polarizations. The beam exiting the interferometer can be subsequently subjected to a suitable polarization transformation. The polarizers can be oriented to preserve or erase the which-way information, thus changing the interference visibility.

A number of quantum erasure experiments with setups different from the abovementioned ones have also been presented [6–8]. Following the experiments proposed in [2, 9–12], we develop an experiment that uses correlated photon pairs to demonstrate quantum erasure in an interferometer. The signal beam traverses the polarization interferometer (PI) and falls on a detector, while the idler beam heralds the detection of single photons, ensuring that we are detecting single photons in the signal beam, and hence observing single-photon interference. We are interested in observing interference patterns in the measured *coincidence* counts for the two beams. It must be kept in mind that the which-path information must be erased to observe interference. The which-path information is switched on or switched off, i.e., is erased by modifying the orientation of a polarizer, which follows the interferometer.

The PI's working is now described.

---

[1] A factor to quantify the extent to which the interference pattern is visible.

7.1.1 The polarization interferometer and quantum erasure

The PI can employ a Mach–Zehnder interferometer with two polarizing beam splitters (figure 7.1a). This is an update to the interferometer shown in figure 2.9 with the beam splitters replaced by polarizing beam splitters. In this scheme, a polarized single photon enters the first polarizing beam splitter PBS_1 and emerges into two possible paths. These paths are recombined at PBS_2, allowing for the possibility of quantum interference. The initial HWP_1 converts an incoming $|H\rangle$ photon to $|D\rangle = (|H\rangle + |V\rangle)/\sqrt{2}$, which is also the state coming out from PBS_2, after propagating through the interferometer. This state renders an equal possibility to register a single count on D_1 or D_2.

An alternative layout is shown in figure 7.1. In this situation, the polarizing beam splitter (PBS) is replaced by another kind of optical element called the beam displacing prism, BDP_1, which provides two output channels for an $|H\rangle$ and a $|V\rangle$ photon, both propagating in the same direction but laterally displaced, in our case by $\approx$4 mm. The two possible paths can be recombined by a second beam displacing prism BDP_2, and out emerges the single photon. This photon can be analyzed through a polarizer oriented at α placed before the single-photon detector.

If $\alpha = 90°$, then a $|V\rangle$c photon strikes the detector, which must have taken the 'up-up' path as shown in figure 7.2a. On the other hand, if $\alpha = 0°$, then a $|H\rangle$ photon strikes the detector and we can be absolutely sure, as depicted in figure 7.2b, that it must have traversed the 'down-down' path. This obtainment of which-path information precludes the possibility of interference. However, if $\alpha = 45°$, then interference can be observed because this setting does not allow us to discern which path the photon has taken.

One method to fetch the interference pattern is to tilt the BDP_2 through a tiny angle about the vertical axis, and observe the intensity falling on the detector. The tilt introduces a phase difference φ between the up-up and down-down paths, allowing one to construct the unitary matrix for the PI

$$\hat{O}_{PI} = \begin{pmatrix} 1 & 0 \\ 0 & e^{i\phi} \end{pmatrix}, \tag{7.1}$$

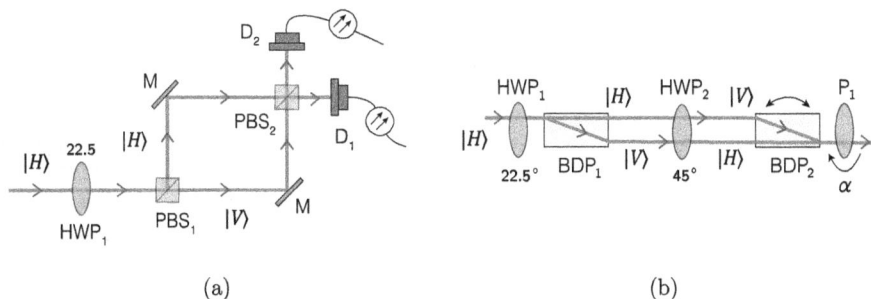

(a)　　　　　　　　　　　　(b)

Figure 7.1. Two variants of a Mach–Zehnder interferometer: (a) using polarizing beam splitters, and (b) using beam displacing prisms.

(a)

(b)

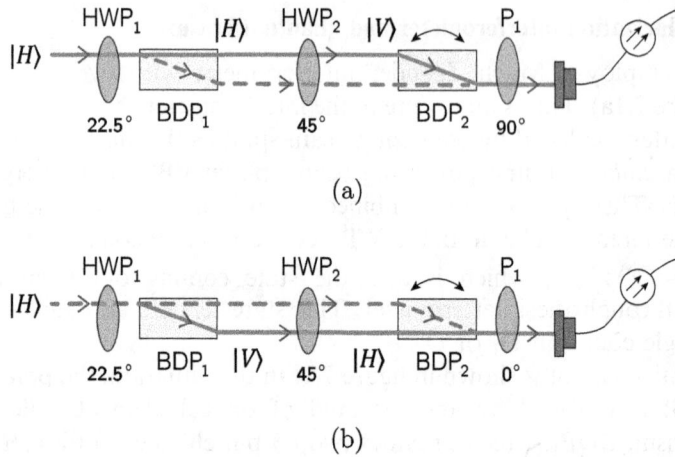

Figure 7.2. Two available paths for a photon: (a) $|H\rangle$ goes through BDP_1 undeviated, changes to $|V\rangle$ after passing through HWP_2, and is displaced by BDP_2; (b) $|V\rangle$ is deflected by BDP_1, turned to $|H\rangle$ by HWP_2, and passes through BDP_2 undeviated.

which acting upon the input state $|D\rangle$ yields

$$|\psi_{\text{out}}\rangle = \hat{O}_{PI}|D\rangle \tag{7.2}$$

$$= \begin{pmatrix} 1 & 0 \\ 0 & e^{i\phi} \end{pmatrix} \frac{1}{\sqrt{2}} \begin{pmatrix} 1 \\ 1 \end{pmatrix} \tag{7.3}$$

$$= \frac{1}{\sqrt{2}} \begin{pmatrix} 1 \\ e^{i\phi} \end{pmatrix} = \frac{1}{\sqrt{2}}(|H\rangle + e^{i\phi}|V\rangle). \tag{7.4}$$

The state $|\psi_{\text{out}}\rangle$ then passes through the polarizer P_1. If the polarizer is oriented at α, then the probability of photodetection is given by

$$P_{|\alpha\rangle} = |\langle \alpha | \psi_{\text{out}}\rangle|^2 = \frac{1}{2}(1 + \sin 2\alpha \cos \phi). \tag{7.5}$$

If the polarizer is set at $\alpha = 45°$, then the probability expression reduces to

$$P_{|45°\rangle} = \frac{1}{2}(1 + \cos \phi), \tag{7.6}$$

which exudes an oscillatory dependence on φ. Hence, an interference pattern can obtained by placing a screen after the polarizer and varying φ. In this interferometer setting, the which-way information has been 'erased', and hence interference fringes are expected. On the contrary, for angles such as $\alpha = 0°$ and $\alpha = 90°$, the probability expression yields

$$P_{|0°\rangle} = P_{|90°\rangle} = \frac{1}{2}, \tag{7.7}$$

and therein the probability of photodetection is independent of the path difference. In other words, changing the path difference will cause no difference in photodetection, i.e., no interference fringes. This is because the which-path information of the signal photon becomes completely available if the photon polarization is known to be either horizontal (in case of $\alpha = 0°$) or vertical (in case of $\alpha = 90°$).

One may wonder what the role of the HWP_2 is given that it is sandwiched between BDP_1 and BDP_2. Its role is to equalize the length differences between the two physical channels inside the interferometer. Had it not been there, the path difference may become larger than the coherence length of the single photons, preventing the appearance of interference fringes. HWP_2 swaps the polarizations in the up-up and down-down paths so that both paths are equalized, making it easier to observe fringes through merely the tilt of a BDP. Please also note that this interferometer takes single photons, albeit from the heralded strategy, and one photon, which is an indivisible quantum, is not *split* into two paths but is rather in a superposition state inside the interferometer. Superposition is one of the most beguiling aspects of quantum physics.

7.1.2 Aligning the interferometer

For experiment Q5, we first need to make sure that detectors A and B are aligned according to the procedure outlined in experiment Q1. After aligning detectors A and B, we need to install the PI in the signal beam path. Note that the Mach–Zehnder interferometer in figure 7.1b causes a displacement in the path of the overall input beam. Therefore, taking into account the displacement of the downconverted beam once the interferometer is installed, detector B should also be displaced. The displacement depends on the specifications of the BDP used. The ones that we use give us a displacement of 4 mm. Alignment on an optical table is highly sensitive. So it might as well happen that while adjusting the new position of detector B, we might end up destroying the whole alignment and have to start all over again. To avoid this, we have outlined a specific procedure to effectively insert the interferometer into the optical setup of experiment Q1 for the purpose of experiment Q5, while taking the necessary precautions of the already aligned setup.

The alignment laser is shone through the detector mount B to the downconversion crystal. If the detectors were properly aligned, this path taken by the alignment laser is the same path that the single photons take while coming to the collection optics. We will start placing components one by one in this path, starting from the far end on the detector side. First, a polarizer P_1 at $0°$ is placed right in front of detector B, which will allow only the H-component of light to pass through. Next, we insert a PI in the setup but with the HWP_2 at $0°$ as shown in figure 7.3a. The BDPs do not displace the H-polarized light, so the beam entering this arrangement of optical elements will pass through undeflected. The hardware is to be placed such that the beam enters BDP_2 and exits BDP_1 on the slightly right-hand side when seen from the location of detector B's mount, as illustrated in figure 7.3a.

(a)

(b)

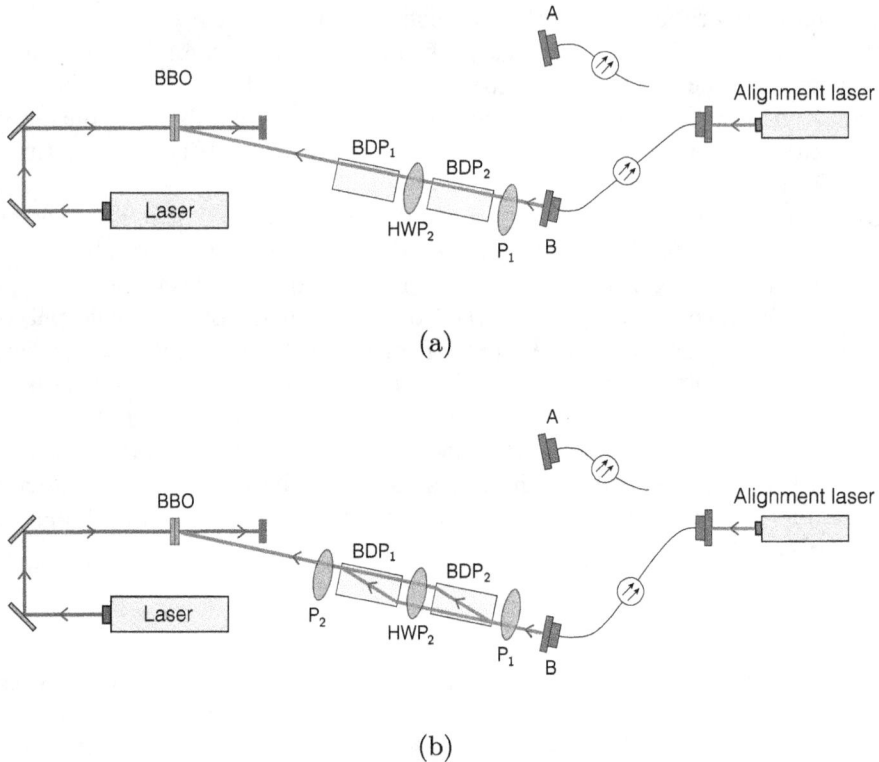

Figure 7.3. Rough alignment of the interferometer. (a) Detector B is aligned, and two irises are placed near the BBO crystals. (b) Detector B is moved 4 mm to the left-hand side (when seen from the detector's side), and the interferometer elements are placed and aligned. The polarizer P_2 is temporary and used for alignment.

By following this procedure, we will have placed the PI in the path that single photons are already taking, i.e., from the BBO crystal to detector B.

Once we turn HWP_2 to $45°$, it will change the polarization of the laser from $|H\rangle$ to $|V\rangle$. This will cause BDP_1 to steer the input light, and it may not exit the BDP_1 at all. To recover the laser beam, detector B needs to be moved to the left by 4 mm. The ideal location of detector B's mount will be such that this displaced beam that eventually exits BDP_1 ultimately hits the BBO crystal. Finally, if we turn the polarizer P_1 to $45°$, then the input beam will split into two portions, take both of the available paths in the PI, and merge as a single beam from BDP_1. This is the situation depicted in figure 7.3b. This step completes the alignment of the PI in the path that the single photons would take when injected from the crystal's side.

Utmost precision is needed to change the tilt of BDP_2. To achieve this, we use a piezo-based actuator. We fix the BDP inside a kinematic mount and replace one of the mount's manual horizontal tilt adjustment screws with the shaft of the piezo motor, as shown in figure 7.4. The actuator had a typical step size of 20 nm. Now, to check if the interferometer is aligned effectively, we set P_1 at $45°$ and place, temporarily, another polarizer, P_2, oriented at $45°$ right after BDP_1. A screen can

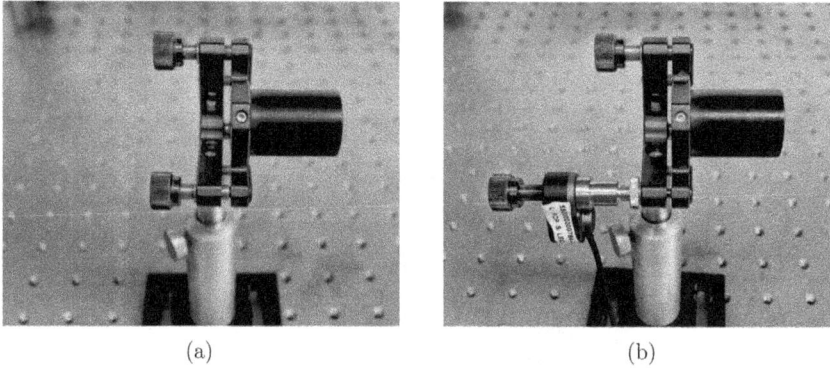

(a)   (b)

Figure 7.4. (a) A BDP mounted on a kinematic mount with a manual adjustment screw. (b) The horizontal tilt knob of the BDP's kinematic mount was replaced with a piezo-based actuator motor.

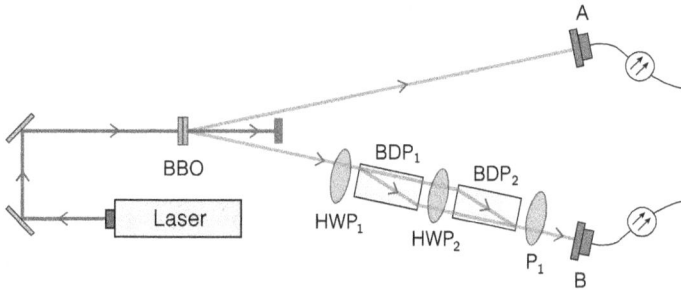

Figure 7.5. Schematic diagram of the single-photon interference and quantum eraser experiment. One of the downconverted photons goes to detector A and heralds the detection of the other photon, which passes through a PI and falls on detector B. The polarizer just before detector B determines the extent to which the which-path information becomes available, controlling the fringe visibility.

be temporarily inserted in front of P_2 to see the interference pattern. When the horizontal tilt of one of the BDPs is changed with the help of the actuator, variation in brightness on the output beam should be observed. These are our interference fringes, which confirm that our interferometer is now functional and ready to go.

It is now time for some final considerations. The beam of photons coming from the BBO is in state $|H\rangle$, but for the beam to take both paths in the interferometer it must be in a superposition, $|D\rangle$. To change the polarization of incoming photons, we insert the half-wave plate (HWP) HWP_1 inside the signal beam. When oriented at $22.5°$, a HWP can turn the $|H\rangle$ state into $|D\rangle$. At the stage, the temporarily used polarizer P_2 is also removed. The alignment laser is removed and the detector B is connected with the APD. The rough alignment is complete and now we need to equalize the interferometer path lengths for the downconverted photons. The experimental scheme and photographs are shown in figures 7.5 and 7.6, respectively.

By turning the lights off and turning the pump beam and the detectors on, a range of the BDP tilt is scanned precisely with the help of the actuator. Oscillation of AB

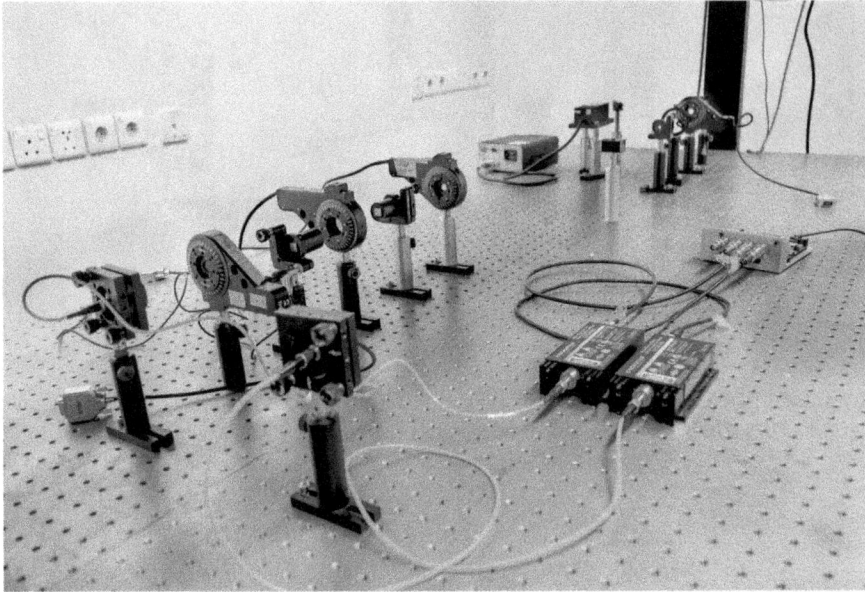

Figure 7.6. Photograph of the single-photon interference and quantum eraser setup, experiment Q5.

coincidence counts is observable when the path difference of the two interferometer arms is within the coherence length of the photons. If the fringes were observable with the classical light, then the single-photon fringes could be found by scanning the actuator step range from $-100\ 000$ to $100\ 000$ steps. The interference fringes will have maximum visibility when the path lengths are equalized, and the visibility may be further improved by very slightly tweaking the tilt of the detector B mount or the rotation angle of the wave plates.

7.1.3 The experiment

Our experiment is based on observing the behavior of single photons when they pass through the PI followed by the information eraser P_1. The idea is that for some polarizer settings, the which-path information is available to us. This is when we know the path that a single photon traverses inside the interferometer, and in such cases we cannot observe any interference. On the contrary, for other settings of P_1, the which-path information becomes unavailable or is 'erased'. This recovers high-visibility interference fringes. The idea is similar to the experiment C3 in chapter 2.

For different orientations (α) of the polarizer, the coincidence counts N_{AB} are recorded against the path difference controlled by the piezo-actuated motor. The coincidence counts are corrected for accidental coincidences and are plotted as a function of the actuator position (figure 7.7). The visibility of the interference fringes is calculated using the following formula [1] and compared with the theoretical prediction:

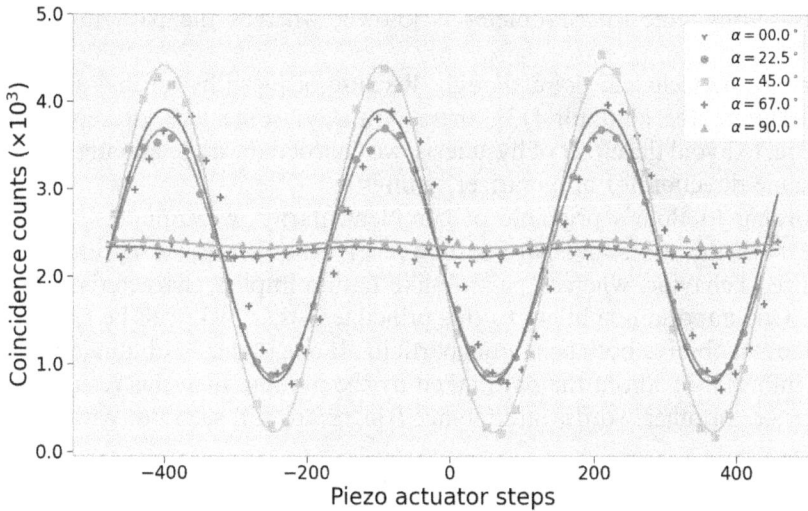

Figure 7.7. A typical variation of detected coincidence counts with respect to change in phase difference for different polarizer orientations. These plots correspond to the first dataset.

Table 7.1. Visibility of interference for different orientations of the polarizer. These results correspond to the first dataset.

Polarizer orientation α (°)	Measured visibility	Predicted visibility
0.0	0.06	0.00
22.5	0.64	0.71
45.0	0.93	1.00
67.5	0.70	0.71
90.0	0.05	0.00

$$V = \frac{I_{\max} - I_{\min}}{I_{\max} + I_{\min}} = \frac{N_{\max} - N_{\min}}{N_{\max} + N_{\min}}, \tag{7.8}$$

where $N_{\max}$ and $N_{\min}$ represent the maximum and minimum number of coincidences.

The visibility can also be predicted using equation (7.5). For the scenarios where the polarizer angle α is neither 0° nor 90°, the maximum and minimum probabilities $N_{\max}$ and $N_{\min}$ are $(1 + \sin 2\alpha)/2$ and $(1 - \sin 2\alpha)/2$, respectively. Inserting these equations into equation (7.8) yields a predicted visibility of $\sin 2\alpha$, which is highest at the polarizer angle of $\alpha = 45°$.

From figure 7.7, it can be seen that the 45° setting has the highest visibility, whereas the 0° and 90° settings produce the lowest visibility. Detailed results are summarized in table 7.1. For the $\alpha = 0°$ and $\alpha = 90°$ settings, the path of the photons through the interferometer is known, whereas the path information is erased for $\alpha = 45°$.

There is one technical point though. Because of piezo hysteresis[2] and the open loop design of the actuator [13], there is always some lack of reproducibility. Therefore, to avoid the effect of hysteresis, we collect data while moving the actuator in the same direction for all polarizer settings.

According to Bohr's principle of complementarity, we cannot simultaneously observe both the wave and particle nature of light. Wave-like nature implies an unlocalized behavior, whereas particle-like nature implies characteristics of localization. One important subtlety to this principle is that there can be circumstances when one can observe both behaviors partially. If one is able to obtain incomplete or partial information about the path taken by the photon, then this results in partial interference. In other words, interference fringes are still seen but with diminished visibility.

The take-home message of this experiment is the bizarre observation that there may be interference even if only a single photon is traversing the interferometer at any instant. This 'wavy' behavior of photons may seem odd because, while doing experiment Q2, we observed the granular nature of light and considered photons to be 'particles' of light and we do not find particles undergoing interference to be reasonably intuitive. Therefore, in light of experiments Q5 and the forthcoming Q+NL, when seen together, one may conclude that although photons may act like particles in *some* ways, one cannot think of them as *classical* particles in the usual sense. There is much more to them. Furthermore, the use of classical semantics is also prone to interpretative danger [14].

7.2 Q+NL: Nonlocal quantum erasure

The nonlocality experiments from chapter 6 were concerned with ascertaining the violation of Bell's inequalities that would jeopardize the tenet of locality, which is considered untouchable in classical physics. A phenomenon that negates those inequalities negates classical physics and a quantum mechanical approach must then be undertaken to study such cases. The experiments employing entangled single-photons clearly violated Bell's inequalities. Similarly, in experiment Q5, we also showed that single photons can interfere, and this interference can be suppressed or recovered by an element placed after any devices occupying the spatial zone where interference can possibly occur.

In this section, we will see that the quantum erasure experiment can also be performed with entangled photon pairs [11, 15]. In such an experiment, the which-path information of a photon can be obtained not only from the signal channel passing through the interferometer but also from the twin idler. This is because the idler and signal photons are entangled.

---

[2] This is the jitter or asymmetry in the displacement of the piezoelectric actuator while moving in opposite directions under identical electrical stimuli.

If the photon path through the interferometer depends on the photon polarization, then a suitable measurement of the idler photon polarization determines the which-path information of the signal photon. Hence, interference fringes are not observed in this case. To observe interference, it must be ensured that the which-path information is erased from *both* the photon beams. This is realized by introducing a polarization measurement strategy with the idler photon. Hence, measurements on the idler photon can determine the erasure of the which-path information of the signal photon.

7.2.1 Erasure with nonlocality

In experiment Q5, we passed single photons through a Mach–Zehnder interferometer and filtered the emanating light through a polarizer. We observed that our choice of the polarizer analysis angle changed the visibility of the interference pattern.

Although the quantum eraser in Q5 gave us an experimental demonstration of wave-particle duality by switching between the wave and particle picture of single photons, Kwiat *et al* [16] argue that such a quantum eraser is not *ideal* because modifying the analysis angle modifies the interferometer itself. In this experiment, QL+NL, we will take an approach similar to [17] and realize quantum erasure using entangled photons. Instead of changing the analysis angle in the signal path, we will send one photon of the entangled pair (the signal) through the interferometer and the other photon (the idler) through a polarization measurement element. Figure 7.8 shows a schematic of this setup. The difference from Q5, therefore, is that the idler now serves *two* purposes. The first purpose is the heralding of the single-photon state in the signal beam. The second purpose is that the measurement of the idler determines the observation of interference in the signal path. We term this experiment 'nonlocal quantum erasure'.

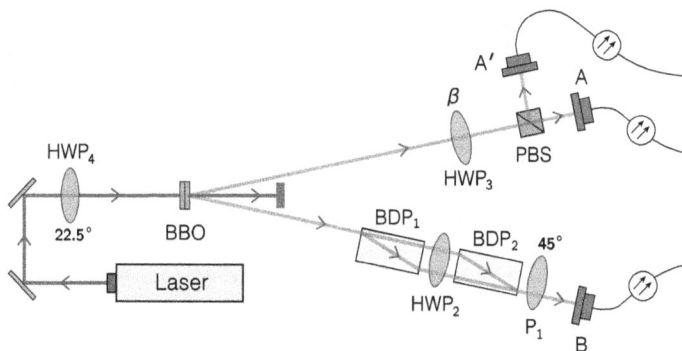

Figure 7.8. A schematic for the experimental settings to perform nonlocal erasure. The interferometer is set in the signal beam's path, while a two-detector measuring apparatus is set in the idler beam's path.

7.2.2 Quantum mechanical prediction for nonlocal erasure

As discussed in chapter 3, a measurement performed on one particle in an entangled state predetermines the output of the other particle. Such correlations are the strongest when we have maximally entangled states, i.e., the Bell states. With type-I downconversion crystals, we can generate two kinds of maximally entangled states,

$$|\Phi^+\rangle = \frac{1}{\sqrt{2}}(|HH\rangle + |VV\rangle),$$

$$|\Phi^-\rangle = \frac{1}{\sqrt{2}}(|HH\rangle - |VV\rangle).$$

If we generate entangled photons in one of these states, then a measurement performed in the idler path in the $\{|H\rangle, |V\rangle\}$ basis will immediately determine the state of the signal beam photons. In other words, the measurement outcome in both paths should be the same, i.e., a $|H\rangle$ measured in the idler path will restrict the measurement outcome in the signal path to yield $|H\rangle$.

Notice that the experimental setup of the interferometer in this setup, shown in figure 7.8, is a bit different from the one we made for experiment Q5 (figure 7.5). HWP_1 has been removed from the setup. In experiment Q5, only one of the crystals in the BBO stack was being used for downconversion, and the state in the signal arm was simply $|H\rangle$, while HWP_1 was set at 22.5° to convert $|H\rangle$ to $|D\rangle$, ensuring that the light entering the interferometer had both $|H\rangle$ and $|V\rangle$ components. Only if the photon is in a superposition of $|H\rangle$ and $|V\rangle$ can it take both paths in the interferometer and form an interference pattern.

In the current experiment, the complete BBO stack produces a superposition state which possesses the ability to take both of the available paths. Hence, HWP_1 is not needed. However, another HWP, HWP_4, is put in the pump beam so that the appropriate Bell state is synthesized after downconversion. This strategy has been adequately described in section 6.3.

Furthermore, unlike experiment Q5, we will not change the polarizer P_1's orientation in this experiment, which is fixed at 45°. Hence, there is no change in the interferometer itself, making this arm of the experiment an implementation of the ideal quantum eraser remaining constant throughout. However, experiment Q +NL differs in the introduction of HWP_3, a PBS, and an additional detector, A', all in the idler path.

If the idler and the signal paths have photons that are not entangled, measuring the state of a photon in the idler beam will not influence the state of the photon in the signal beam, and hence no change in the interference pattern should be observed. The situation radically changes for entangled photon pairs.

7.2.3 The experiment

We can generate the entangled state as outlined in section 6.3, set the interferometer as detailed in section 7.1.2, and align the PBS in the idler path using the double back-propagation technique from section 4.3.5. With these preliminaries dealt with, we can perform the experiment.

As we saw in Q5 that interference fringes are obtained by changing the length of the two paths taken by the light in the interferometer. We do this by tilting BDP_2 using the piezoelectric actuator. The measurement of a photon's polarization state determines its path, resulting in modulating the visibility of the interference fringes. In experiment Q5, we determined the path by changing the analysis angle, α, at P_1 at the end of the interferometer. We obtained maximum visibility for analysis angle $45°$, and minimum visibility for analysis angles $0°$ and $90°$. Since we have a different measurement criterion in this experiment, we will fix this polarizer's angle at $45°$, which, if unconnected with the idler photonic beam, furnishes maximum fringe visibility.

Since the two beams are entangled, a measurement in the idler path should also provide us with the path information of the signal photon, washing away the fringes in the signal path. Given that our interferometer's two paths distinguish between $|H\rangle$ and $|V\rangle$ polarized photons, determining either of the states will give complete 'which-way' information for the photon, and hence destroy the interference pattern. For example, if we set the measurement basis in the idler path to $\{|H\rangle, |V\rangle\}$, then we will obtain complete information of the path and we observe no fringes. On the other hand, if the measurement basis is set to $\{|D\rangle, |A\rangle\}$ or to $\{|L\rangle, |R\rangle\}$, then we will not be able to determine the path that the photon takes and the fringes should be at maximum visibility. Measurement in $\{|H\rangle, |V\rangle\}$ basis can be performed by orienting HWP_3 at $0°$, and measurement in $\{|D\rangle, |A\rangle\}$ basis can be performed by orienting HWP_3 at $22.5°$. This is in accordance with the settings presented in the top two rows of table 5.1. Note that both the $\{|D\rangle, |A\rangle\}$ and the $\{|L\rangle, |R\rangle\}$ bases are completely unbiased with respect to the $\{|H\rangle, |V\rangle\}$ basis. We can use either.

Setting the stage more clearly, we gradually change HWP_3's orientation from $0°$ to $22.5°$ to $45°$, i.e., from $\{|H\rangle, |V\rangle\}$ to $\{|D\rangle, |A\rangle\}$ basis, and then back to $\{|H\rangle, |V\rangle\}$. The experiment can be neatly analyzed with the help of quantum state progression. The entanglement generation scheme yields the Bell state,

$$|\Phi^+\rangle = \frac{1}{\sqrt{2}}(|HH\rangle + |VV\rangle), \tag{7.9}$$

where the first state goes to the idler and the second to the signal. Only the PI in the signal path transforms the state under the transformation $\hat{1} \otimes \hat{O}_{PI}$ where $\hat{O}_{PI}$ is given by equation (7.1),

$$|\Phi^+\rangle \xrightarrow{\hat{1} \otimes \hat{O}_{PI}} \frac{1}{\sqrt{2}}\left(|HH\rangle + e^{i\phi}|VV\rangle\right) = |\psi_1\rangle. \tag{7.10}$$

We set HWP_3 at an angle β, which allows us to apply the transformation $\hat{J}_{HWP}(\beta) \otimes \hat{1}$, which can be picked from table 2.2, resulting in the correlated state,

$$|\psi_1\rangle \xrightarrow{\hat{J}_{HWP}(\beta) \otimes \hat{1}} \frac{1}{\sqrt{2}}\Big(\cos(2\beta)|HH\rangle + e^{i\phi}\sin(2\beta)|HV\rangle$$

$$+ \sin(2\beta)|VH\rangle - e^{i\phi}\cos(2\beta)|VV\rangle\Big) = |\psi_{\text{out}}\rangle. \tag{7.11}$$

This is followed by a measurement. The A and A' channels on the idler project onto $|V\rangle$ and $|H\rangle$, respectively, whereas in the signal channel, the P_1 oriented at $45°$ projects on the $|D\rangle$ state. This allows us to calculate the probabilities,

$$P_{|H\rangle\otimes|D\rangle} = |(\langle H| \otimes \langle D|)|\psi_{\text{out}}\rangle|^2 = \frac{1}{4}(1 + \cos\phi\,\sin(4\beta)), \tag{7.12}$$

$$P_{|V\rangle\otimes|D\rangle} = |(\langle V| \otimes \langle D|)|\psi_{\text{out}}\rangle|^2 = \frac{1}{4}(1 - \cos\phi\,\sin(4\beta)). \tag{7.13}$$

The probabilities preserve the ϕ dependence and show interference unless $\beta = 0$ or $\beta = \pi/4$. Interference is maximized when $\beta = \pi/8$ which corresponds to the idler photon being measured in the $\{|D\rangle, |A\rangle\}$ basis. This simple calculation indicates that if $\beta = 0$ or $\beta = \pi/4$ (measurement in $\{|H\rangle, |V\rangle\}$ basis), we can tell which path the photon takes in the entangled pair, deleting fringes and providing which-path information. To put it mildly, this is beautiful! A photon's measurement in a distant location determines whether we can see another photon interfering or not. This is why this observation is called nonlocal erasure.

Another point is noteworthy. In experiment Q+NL, the polarizer P_1 is not adjusted. Hence, the PI is kept totally unchanged as we observe or erase which-path information. All the activity happens in the correlated idler beam; a change in measurement orientation modifies the interference in the signal beam.

For each orientation, we also measure the visibility using the same formula as before,

$$V = \frac{N_{\text{max}} - N_{\text{min}}}{N_{\text{max}} + N_{\text{min}}}. \tag{7.14}$$

The theoretical visibility can be predicted from equations (7.12) and (7.13). For example, if we are looking at the N_{AB} coincidences, then we have (for $\beta \leqslant \pi/4$),

$$V = \frac{P_{|V\rangle\otimes|D\rangle}\,|_{\text{max}} - P_{|V\rangle\otimes|D\rangle}\,|_{\text{min}}}{P_{|V\rangle\otimes|D\rangle}\,|_{\text{max}} + P_{|V\rangle\otimes|D\rangle}\,|_{\text{min}}} = \sin(4\beta). \tag{7.15}$$

We can choose either N_{AB} or $N_{AB'}$ counts to calculate the experimental visibilities. The results summarized in figure 7.9 show the effect of determining the polarization state in the idler beam on the visibility of fringes in the signal beam. We see maximum visibility when the measurement basis is $\{|D\rangle, |A\rangle\}$, and we have no which-path information, whereas for $\{|H\rangle, |V\rangle\}$ basis, we get the least visible fringes. The visibilities are also compared with theoretical predictions in table 7.2.

The results shown in figure 7.9 are for the entangled state $|\Phi^+\rangle$. On the other hand, if we had a separable state, then a measurement in the idler path should not have any

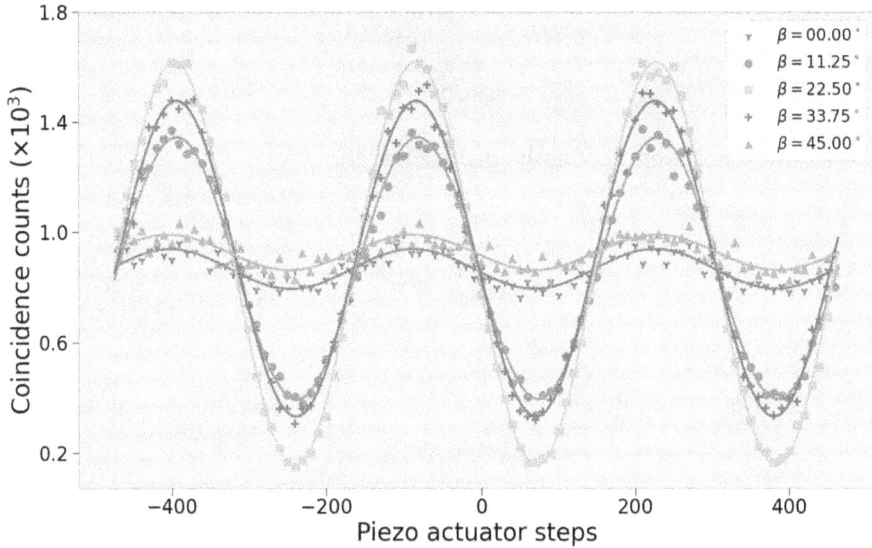

Figure 7.9. Graph for comparing visibility with changing measurement basis angle in the idler beam while measuring an entangled state. The bases are set up by the HWP_3 angle β. The graphs have been adjusted on the horizontal axis to position them for better data representation.

Table 7.2. Predicted and measured visibility of interference for different orientations of HWP_3 for the entangled state.

HWP_3 orientation β (°)	Measured visibility	Predicted visibility
0.00	0.13	0.00
11.25	0.61	0.71
22.50	0.84	1.00
33.75	0.67	0.71
45.00	0.12	0.00

effect on the state of photons in the signal path. A simple separable state can be generated by turning HWP_4 at $0°$ and reintroducing HWP_1 in the signal path at $22.5°$. In this way, the BBO pair will downconvert the single photons into state $|HH\rangle$ and HWP_1 in the signal path will turn the signal beam photons to $|D\rangle$ state. Hence, the combined two-qubit state will be,

$$|\psi\rangle = |H\rangle \otimes \frac{1}{\sqrt{2}}(|H\rangle + |V\rangle) = \frac{1}{\sqrt{2}}(|HH\rangle + |HV\rangle). \tag{7.16}$$

Repeating the process that lead to equations (7.10) and (7.11), just prior to detection, the separable state propagates to

$$\frac{1}{\sqrt{2}}(\cos(2\beta)|HH\rangle + e^{i\phi}\cos(2\beta)|HV\rangle +$$

$$\sin(2\beta)|VH\rangle + e^{i\phi}\sin(2\beta)|VV\rangle) = |\psi_{out}^{(sep)}\rangle, \tag{7.17}$$

from which the probabilities can be deduced as follows,

$$P_{|H\rangle \otimes |D\rangle} = \left| (\langle H| \otimes \langle D|)|\psi_{out}^{(sep)}\rangle \right|^2$$
$$= \cos^2(2\beta)\cos^2\left(\frac{\phi}{2}\right), \tag{7.18}$$

$$P_{|V\rangle \otimes |D\rangle} = \left| (\langle V| \otimes \langle D|)|\psi_{out}^{(sep)}\rangle \right|^2$$
$$= \sin^2(2\beta)\cos^2\left(\frac{\phi}{2}\right), \tag{7.19}$$

which corresponds to counts $N_{A'B}$ and N_{AB}, respectively.

This visibility for the separable state can also be calculated using either $N_{A'B}$ or N_{AB} counts, but it is to be noted that when HWP$_3$ is at $0°$, detector A' will get maximum counts while A will get minimum counts, and the situation will reverse if HWP$_3$ is at $45°$. So, for better comparison of visibility, it is recommended to switch the observing detector (from A' to A) once HWP$_3$'s angle goes above $22.5°$. The results are shown in figure 7.10.

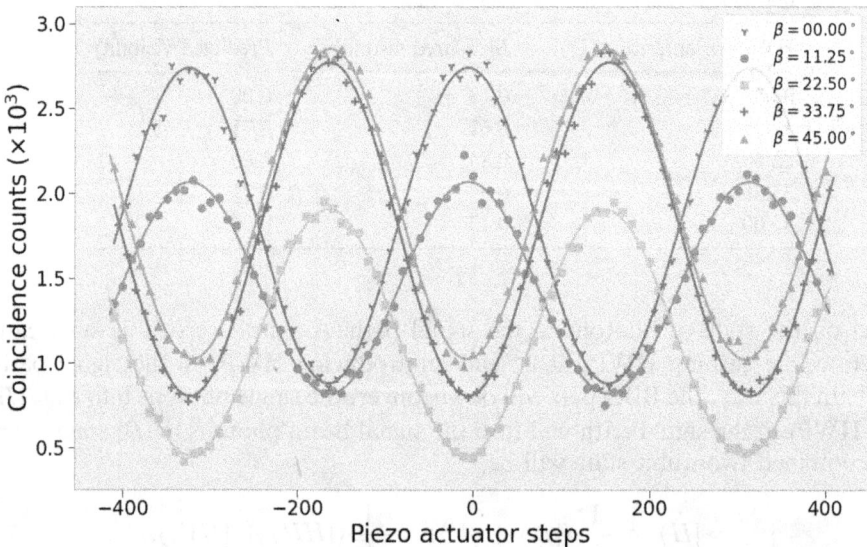

Figure 7.10. Graph for comparing visibility with changing measurement basis angle in the idler beam while measuring a separable state. The bases are set up by the HWP$_3$ angle β. The observing detector was switched for HWP$_3$ angles $37.5°$ and $45°$, and the graphs have been adjusted horizontally to position them for better data representation.

Table 7.3. Predicted and measured visibility of interference for different orientations of HWP$_3$ for the separable state.

HWP$_3$ orientation β (°)	Measured visibility	Predicted visibility
0.00	0.54	1.00
11.25	0.50	1.00
22.50	0.64	1.00
33.75	0.57	1.00
45.00	0.49	1.00

As we can see, the visibility in this case has not changed significantly. Theoretically, it remains close to the value one, whereas experimental visibilities are also shown in table 7.3. From the probabilities calculated, the maximum amplitude of a fringe is $\cos^2 2\beta$ ($\sin^2 2\beta$) for $N_{AB'}$ (N_{AB}) and the minimum in each case is zero, resulting in $V = 1$. This shows that the measurement of the idler beam, in this case, has no effect on the state of photons in the signal beam and the visibility is not affected. There is a clear distinction in the observed results between an entangled and a separable state.

References

[1] Galvez E J, Holbrow C H, Pysher M J, Martin J W, Courtemanche N, Heilig L and Spencer J 2005 Interference with correlated photons: five quantum mechanics experiments for undergraduates *Am. J. Phys.* **73** 127–40

[2] Beck M 2012 *Quantum Mechanics: Theory and Experiment* (Oxford: Oxford University Press)

[3] Schwindt P D D, Kwiat P G and Englert B-G 1999 Quantitative wave-particle duality and nonerasing quantum erasure *Phys. Rev.* A **60** 4285

[4] Maries A, Sayer R and Singh C 2020 Can students apply the concept of 'which-path' information learned in the context of Mach-Zehnder interferometer to the double-slit experiment? *Am. J. Phys.* **88** 542–50

[5] Schneider M B and LaPuma I A 2002 A simple experiment for discussion of quantum interference and which-way measurement *Am. J. Phys.* **70** 266–71

[6] Herzog T J, Kwiat P G, Weinfurter H and Zeilinger A 1995 Complementarity and the quantum eraser *Phys. Rev. Lett.* **75** 3034

[7] Hong C K and Noh T G 1998 Two-photon double-slit interference experiment *J. Opt. Soc. Am.* B **15** 1192–7

[8] Kim Y-H, Yu R, Kulik S P, Shih Y and Scully M O 2000 Delayed 'choice' quantum eraser *Phys. Rev. Lett.* **84** 1

[9] Kwiat P G and Englert B-G 2004 Quantum erasing the nature of reality or, perhaps, the reality of nature *Science and Ultimate Reality: Quantum Theory, Cosmology, and Complexity* (Cambridge: Cambridge University Press) p 306

[10] Gogo A, Snyder W D and Beck M 2005 Comparing quantum and classical correlations in a quantum eraser *Phys. Rev.* A **71** 052103

[11] Ashby J M, Schwarz P D and Schlosshauer M 2016 Delayed-choice quantum eraser for the undergraduate laboratory *Am. J. Phys.* **84** 95–105

[12] DiBrita N S and Galvez E J 2023 An easier-to-align Hong-Ou-Mandel interference demonstration *Am. J. Phys.* **91** 307–15

[13] Thorlabs 2022 Pia Series Piezo Inertia Actuators User Guide. edition: Etn017852-d02

[14] Englert B-G 1996 Fringe visibility and which-way information: an inequality *Phys. Rev. Lett.* **77** 2154–7

[15] Walborn S P, Terra Cunha M O, Pádua S and Monken C H 2002 Double-slit quantum eraser *Phys. Rev. A* **65** 033818

[16] Kwiat P G, Steinberg A M and Chiao R Y 1994 Three proposed "quantum erasers" *Phys. Rev. A* **49** 61–8

[17] Gogo A, Snyder W D and Beck M 2005 Comparing quantum and classical correlations in a quantum eraser *Phys. Rev. A* **71** 052103

IOP Publishing

Quantum Mechanics in the Single-Photon Laboratory
(Second Edition)

Muhammad Sabieh Anwar, Faizan-e-Ilahi, Syed Bilal Hyder and Muhammad Hamza Waseem

Chapter 8

Quantum state tomography

In the previous experiments Q3, Q4, Q5, and Q+NL, we used one of the two downconverted photons as a herald for detection of the other photon. Our focus, therefore, was mainly on studying the quantum state of the signal photon. However, the complete study of entangled states requires the quantum state of *both* the photons forming the entangled pair to be investigated. In experiments NL1–NL3, we also saw that quantum mechanics explains certain two-photon polarization states which violate the apparently self-evident assumptions of locality. This pair of photons was *entangled* in polarization.

In this chapter, we extend the ideas of chapter 5 and attempt to measure the *full* quantum state of the pair of downconverted photons, expressed in terms of the polarization degree of freedom. This is achieved through the technique of quantum state tomography (QST).

QST is a fancy name for estimating the quantum state. It is achieved by using a number of carefully orchestrated measurements. There are a few caveats, though. As established in experiments Q3 and Q4, performing measurement of a quantum particle perturbs its state. Therefore, QST cannot be used to unambiguously determine the state of a *single* particle in just one go. Rather, it is performed systematically in successive stages on many identical copies of the quantum state and since there can never be an infinite supply of particles, so we can only make educated best guesses about the state. In all cases, we need to devise an algorithmic procedure for state estimation.

We have seen in chapter 2 that the Stokes parameters S_0, S_1, S_2, and S_3 describe the polarization state of light [1]. These parameters completely characterize the state. One can, therefore, imagine that state tomography implies the measurement of precisely these parameters. This is indeed true.

We have also learned that certain light beams can be seen as ensembles of two-level quantum systems (comprising photons), with the two polarization degrees of freedom defining the quantum state. The polarization state is completely and most generally described not by a ket or vector but by a density matrix of the form

$$\hat{\rho} = \begin{pmatrix} a & b + ic \\ b - ic & 1 - a \end{pmatrix} \tag{8.1}$$

where a, b, and c are real numbers making $\hat{\rho}$ into a unit trace and Hermitian, positive semi-definite matrix[1]. Chapter 5 dealt with a pure quantum state, this chapter deals with the density matrix.

The matrix contains three real parameters, the a, b, and c that are to be estimated. Apparently, this contradicts the need for specifying four Stokes parameters. This discrepancy is resolved by noting that the parameter S_0 depends on the intensity in the classical case and the number of counts in the quantum mechanical case. Measuring for longer times over higher intensity beams will increase S_0, which does not really affect the polarization state but merely distinguishes how *big* the observed signal is. Therefore, one of the Stokes parameters is determining the signal-to-noise ratio, and it is only the trio S_1, S_2, and S_3 normalized against S_0 which characterizes the state. Therefore, there should exist a direct one-to-one mapping between the Stokes parameters and the real parameters a, b, and c that describe the density matrix. This chapter will build upon this understanding.

There exist simple tomographic techniques in which experimental data are linearly transformed to find the density matrix of a quantum state. However, because of experimental noise, these methods may sometimes not fetch density matrices corresponding to realizable physical states describable by Hermitian, positive semi-definite matrices of unit trace of the form (8.1). In fact, many experimentally measured matrices usually fail to fulfill the positive semi-definiteness condition [2]. As a workaround, a maximum-likelihood estimation approach has been proposed and successfully utilized [3–7]. According to this method, the density matrix, which has the maximum likelihood of having generated the measured dataset, is obtained via numerical optimization, and the semi-definite positivity condition is hard-wired into the optimization routine. This chapter's discussions closely follow reference [8], which is an excellent introduction to using maximum-likelihood techniques for estimating the polarization states of photons.

A little bit of a survey here. Tomographic methods have been successfully used for measurement in quantum mechanical systems of a vast variety and complexity [9–15]. Particularly, these methods have worked remarkably well in studying the quantum state of qubits based on polarization-entangled pairs of photons generated through downconversion [16]. Our focus will be on the QST of such systems, but the discussion is applicable to other two-level quantum states as well.

In this chapter, we will extend the discussion of techniques for QST, already initiated in experiments Q3 and Q4, in the context of correlated two-level quantum optical systems or qubits. We will particularly focus on two techniques, namely a linear tomographic reconstruction and a maximum-likelihood technique.

---

[1] Density matrices have been introduced in section 3.6.

The former technique linearly relates the density matrix to a series of measurements and is important for understanding the theoretical basis of tomography using idealized measurements. However, this method does not always return valid density matrices for physical systems. On the other hand, maximum likelihood is based on numerical optimization and is an adaptation of the former technique to real, nonideal systems. Hence, it also returns valid density matrices for imperfect experimental conditions. This is the method that we have employed for estimating the quantum state in our experiments. To begin our discussion, we will revisit the relationship between the Stokes parameters and QST.

8.1 Qubits, Stokes parameters, and tomography

As the introduction mentions, an analogy exists between measuring the polarization state of light and estimating the density matrix of two-level quantum systems. This analogy is the subject of this section. Before discussing tomography for two qubits, we will investigate the single-qubit case. This will help us to develop an intuitive picture of quantum state representation and tomography before we can investigate the extended case of two qubits. In essence, we have covered many of these aspects in earlier parts of the book, but we think it would be worthwhile gathering all the bits and pieces together in one place, *one more time*.

8.1.1 The Bloch sphere for pure states

As we saw in chapter 3, we can generally express any single qubit by equation (3.7) if it is a pure state. This representation is not only sufficient for describing a pure state but also allows us to describe the action of an operator by straightforward matrix manipulation. For example, projection or unitary operation on the pure state $|\psi\rangle$ can be described by $\hat{P}_\phi|\psi\rangle = |\phi\rangle\langle\phi|\psi\rangle$ and $\hat{U}|\psi\rangle$, respectively.

Figure 8.1(a) shows a pure state $|\psi\rangle$ lying on the surface of the Bloch sphere first introduced in figure 2.4(a). The sphere identifies three axes as pointing along the pure states $|D\rangle$, $|L\rangle$, and $|H\rangle$. Earlier, we have also used different labels for these states:

$$|H\rangle \equiv |0\rangle \tag{8.2}$$

$$|L\rangle \equiv \frac{1}{\sqrt{2}}(|0\rangle + i|1\rangle), \quad \text{and} \tag{8.3}$$

$$|D\rangle \equiv \frac{1}{\sqrt{2}}(|0\rangle + |1\rangle). \tag{8.4}$$

Each axis represents one choice of the basis vectors. For example the axis pointing in the $|D\rangle$ direction corresponds to the $\{|D\rangle, |A\rangle\}$ basis, $|L\rangle$ defines the $\{|L\rangle, |R\rangle\}$ basis, and finally $|H\rangle$ defines the orientation of the $\{|H\rangle, |V\rangle\}$ basis. Note that quantum mechanically, $|H\rangle$ and $|D\rangle$ are not orthogonal ($\langle D|H\rangle = \frac{1}{\sqrt{2}}$) but in the Bloch sphere picture these states may appear to lie along orthogonal axes in the Cartesian sense.

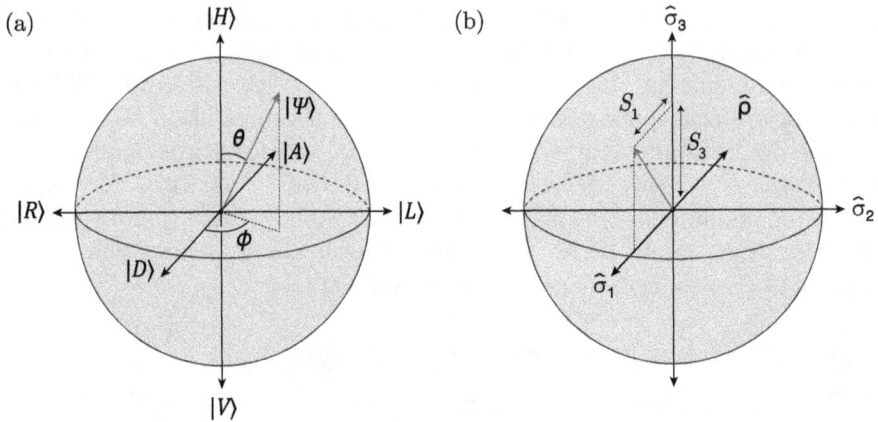

Figure 8.1. (a) Any pure qubit state $|\psi\rangle$ can be represented on the Bloch sphere. This figure is identical to figure 2.4, merely with a relabeling of axes. (b) Another picture is useful when we consider general mixed states. This is also the Bloch sphere with the axes relabeled as the three zero-trace Pauli operators. An arbitrary density matrix ρ is shown by the thick circle with projections S_1 and S_3 on $\hat{\sigma}_1$ and $\hat{\sigma}_3$, respectively.

Quantum orthogonality does not mean orthogonality in the Bloch sphere representation.

The pure state $|\psi\rangle$ is a point on the surface, and its position is characterized by angles θ and φ. In fact, these angles act as parameters that map any pure state onto the surface of the sphere and are shown in figure 8.1. These values are the polar coordinates of the pure state they represent. Apart from the representation of a single-qubit state, the Bloch sphere also serves another purpose—it can be used to represent a unitary operation. On the Bloch sphere, any unitary operation is mapped to a rotation about an axis. We have seen that wave plates are used to perform unitary operations, and hence in the Bloch sphere picture act as rotations about some axis. The axis of rotation is affixed by the orientation of the fast axis of the wave plate and the amount of rotation by the kind of wave plate, i.e., half-wave plate (HWP) or full-wave plate.

8.1.2 The Bloch sphere for density matrices

If the quantum state is mixed, then we need to resort to density matrices, which can generally be expressed as

$$\hat{\rho} = \sum_i p_i |\psi_i\rangle\langle\psi_i| \tag{8.5}$$

$$= \begin{pmatrix} A & Ce^{i\phi} \\ Ce^{-i\phi} & B \end{pmatrix}, \tag{8.6}$$

which is just another way of writing down the matrix in equation (8.1). Here A, B, and C are real, non-negative numbers and are related as $A + B = 1$ and $C = \sqrt{AB}$. Furthermore, φ is a real number, positive or negative. The coefficient p_i denotes the probabilistic weighting ($\sum_i p_i = 1$) [17]. Completely determining this matrix requires

$4 - 1 = 3$ parameters (since $A + B = 1$). The relationship between these parameters and the $((a, b, c))$ from equation (8.1) is as follows,

$$A = a, \tag{8.7}$$

$$B = 1 - A = 1 - a, \tag{8.8}$$

$$b = C \cos \phi, \tag{8.9}$$

$$c = C \sin \phi. \tag{8.10}$$

Using the definitions of the canonical polarization states spelt out in table 3.1, we can construct the so-called Pauli spin operators as

$$\hat{\sigma}_1 \equiv |D\rangle\langle D| - |A\rangle\langle A| = \begin{pmatrix} 0 & 1 \\ 1 & 0 \end{pmatrix}, \tag{8.11}$$

$$\hat{\sigma}_2 \equiv |L\rangle\langle L| - |R\rangle\langle R| = \begin{pmatrix} 0 & -i \\ i & 0 \end{pmatrix}, \tag{8.12}$$

$$\hat{\sigma}_3 \equiv |H\rangle\langle H| - |V\rangle\langle V| = \begin{pmatrix} 1 & 0 \\ 0 & -1 \end{pmatrix} \tag{8.13}$$

and the 2×2 identity matrix

$$\hat{\sigma}_0 \equiv |H\rangle\langle H| + |V\rangle\langle V| = \begin{pmatrix} 1 & 0 \\ 0 & 1 \end{pmatrix}. \tag{8.14}$$

These operators $\{\hat{\sigma}_0, \hat{\sigma}_1, \hat{\sigma}_2, \hat{\sigma}_3\}$ form a complete basis for 2×2 Hermitian matrices, and hence the density matrix in equation (8.6) can also be expressed as

$$\hat{\rho} = \frac{1}{2} \sum_{i=0}^{3} S_i \hat{\sigma}_i, \tag{8.15}$$

where $(S_1, S_2, S_3)^T$ is sometimes called the *Bloch vector* or the *Stokes vector*. The coefficients are simply the Stokes parameters for light. We will show this in just a moment. Due to the unit trace condition, $S_0 = 1$ always tags along for convenience.

A more general view of the Bloch sphere is illustrated in figure 8.1(b), which relabels the axes as the Pauli operators. This conceptual diagram allows us to represent mixed states in addition to pure states. Consider a quantum state (8.15), which lives somewhere on the Bloch sphere. In chapter 3, we have learned that for a pure state $\hat{\rho}^2 = \hat{\rho}$, so we have $Tr(\hat{\rho}^2) = 1$. Using the representation of the density matrix above, we have

$$\hat{\rho}^2 = \frac{1}{4} \left(\sum_{i=0}^{3} S_i \hat{\sigma}_i \right) \left(\sum_{k=0}^{3} S_k \hat{\sigma}_k \right) \tag{8.16}$$

$$= \frac{1}{4} \sum_{i,k=0}^{3} S_i S_k \hat{\sigma}_i \hat{\sigma}_k \tag{8.17}$$

which leads to (using $S_0 = 1$)

$$Tr(\hat{\rho}^2) = \frac{1}{2}\left(1 + S_1^2 + S_2^2 + S_3^2\right). \tag{8.18}$$

Therefore, pure states have $S_1^2 + S_2^2 + S_3^2 = 1$ and for mixed states, $S_1^2 + S_2^2 + S_3^2 < 1$ with the extreme case $S_1^2 + S_2^2 + S_3^2 = 0$ denoting the maximally mixed state.

This discussion reveals an interesting geometrical interpretation of the S coefficients. The mixed state $\hat{\rho}$ in figure 8.1(b) is shown to lie inside the Bloch sphere, which is now shown with the relabeled axes. Suppose that the state is represented by the solid circle and lies wholly in the $\hat{\sigma}_1$–$\hat{\sigma}_3$ plane. If we were to drop a projection of the state onto the $\hat{\sigma}_1$ axis, then we can measure the distance from the origin. This distance is measured to be S_1. Similarly, if the projection is taken along the $\hat{\sigma}_2$ axis, the distance is S_2, while for our particular chosen state $S_2 = 0$. Finally, the distance of the projected point along the $\hat{\sigma}_3$ axis is S_3. Each of these S parameters can be positive or negative—positive when the projection is on one side of the origin and negative when on the diametrically opposite side. We now motivate why these coefficients are identical to the familiar Stokes parameters, described in section 2.2.4.

8.1.3 Stokes parameters as state projections on the Bloch sphere

The Stokes parameters, which provide a complete description of the polarization state of light, are classically defined in terms of intensity measurements in the $\{|D\rangle,$ $|A\rangle\}$, $\{|L\rangle, |R\rangle\}$ and $\{|H\rangle, |V\rangle\}$ bases [18]. For single photons, the intensities can be replaced by numbers of photodetections, or probabilities which are directly proportional to photocounts. If the basis states are given by $\{|D\rangle, |A\rangle\}$, $\{|L\rangle, |R\rangle\}$ and $\{|H\rangle, |V\rangle\}$ representing the canonical polarization states (see table 3.1 in chapter 3), then the Stokes parameters are defined as in equations (5.42)

$$
\begin{aligned}
S_0 &= P_{|H\rangle} + P_{|V\rangle}, \\
S_1 &= P_{|D\rangle} - P_{|A\rangle}, \\
S_2 &= P_{|L\rangle} - P_{|R\rangle}, \text{ and} \\
S_3 &= P_{|H\rangle} - P_{|V\rangle},
\end{aligned}
\tag{8.19}
$$

where $P_{|\phi\rangle}$ is the probability of measuring the photon in the polarization state $|\phi\rangle$. In chapter 2, which dealt with a purely classical description of light, we defined the Stokes parameters as,

$$
\begin{aligned}
S_0 &= I_H + I_V, \\
S_1 &= I_D - I_A, \\
S_2 &= I_L - I_R, \text{ and} \\
S_3 &= I_H - I_V,
\end{aligned}
\tag{8.20}
$$

with the I_H representing the intensity of detecting the light with horizontal (H) polarization etc. Comparing the eight expressions from equations (8.19) and (8.20), we see that the quantum version of Stokes parameters is inferred from probabilities whereas in the classical picture they are inferred from intensities. In quantum experiments, probabilities are deduced from normalized counts, whereas in classical experiments intensities arise out of photocurrents spewed out by detectors.

It is also important to remember that for a state described by the density matrix $\hat{\rho}$, $P_{|\phi\rangle}$ is given by

$$P_{|\phi\rangle} = \langle\phi|\hat{\rho}|\phi\rangle = Tr\{|\phi\rangle\langle\phi|\hat{\rho}\}. \tag{8.21}$$

Using equations (8.6) and (8.21), we have

$$
\begin{aligned}
P_{|\phi\rangle} &= \langle\phi|\left(\sum_i p_i|\psi_i\rangle\langle\psi_i|\right)|\phi\rangle \\
&= \sum_i p_i\langle\phi|\psi_i\rangle\langle\psi_i|\phi\rangle \\
&= \sum_i p_i|\langle\phi|\psi_i\rangle|^2
\end{aligned} \tag{8.22}
$$

where each term $|\langle\phi|\psi_i\rangle|^2$ represents the probability of projecting state $|\psi_i\rangle$ onto $|\phi\rangle$. These terms are mixed up the weighting factor $_{pi}$, which shows what fraction of $|\psi_i\rangle\langle\psi_i|$ contributes to the statistical mixture $\hat{\rho}$.

Using the definition of the state (8.15) and how we extract probabilities from it, equation (8.21), as well as the definitions of the Pauli operators in equations (8.11)–(8.14), we obtain the series of straightforward results

$$S_1 = Tr(\hat{\rho}\hat{\sigma}_1) = Tr(\hat{\rho}(|D\rangle\langle D| - |A\rangle\langle A|)) = P_{|D\rangle} - P_{|A\rangle} \tag{8.23}$$

$$S_2 = Tr(\hat{\rho}\hat{\sigma}_2) = Tr(\hat{\rho}(|L\rangle\langle L| - |R\rangle\langle R|)) = P_{|L\rangle} - P_{|R\rangle} \tag{8.24}$$

$$S_3 = Tr(\hat{\rho}\hat{\sigma}_3) = Tr(\hat{\rho}(|H\rangle\langle H| - |V\rangle\langle V|)) = P_{|H\rangle} - P_{|V\rangle}. \tag{8.25}$$

It would be worth deriving one of these expressions, say (8.23). We start with the measurement outcome $Tr(\hat{\rho}\hat{\sigma}_1)$ and show that this equals S_1.

First, we resolve $\hat{sigma}_1$ in accordance with equation (8.11), yielding

$$Tr(\hat{\rho}\hat{\sigma}_1) = Tr(\hat{\rho}|D\rangle\langle D| - \hat{\rho}|A\rangle\langle A|), \tag{8.26}$$

and from the definition of trace, equation (8.21), we obtain,

$$\langle D|\hat{\rho}|D\rangle - \langle A|\hat{\rho}|A\rangle. \tag{8.27}$$

Finally, we can insert the general form of the density operator from equation (8.15) producing,

$$\frac{1}{2}\langle D|\sum_{i=0}^{3}S_i\hat{\rho}_i|D\rangle - \frac{1}{2}\langle A|\sum_{i=0}^{3}S_i\hat{\rho}_i|A\rangle = \frac{1}{2}\sum_{i=0}^{3}S_i\left(\langle D|\hat{\rho}_i|D\rangle - \frac{1}{2}\langle A|\hat{\rho}_i|A\rangle\right). \tag{8.28}$$

From the resolution of the Pauli operators, we can obtain the following results,

$$\langle D|\hat{\sigma}_0|D\rangle - \langle A|\hat{\sigma}_0|A\rangle = 0,$$
$$\langle D|\hat{\sigma}_1|D\rangle - \langle A|\hat{\sigma}_1|A\rangle = 2,$$
$$\langle D|\hat{\sigma}_2|D\rangle - \langle A|\hat{\sigma}_2|A\rangle = 0, \quad \text{and} \tag{8.29}$$
$$\langle D|\hat{\sigma}_3|D\rangle - \langle A|\hat{\sigma}_3|A\rangle = 0,$$

which shows that the Expression 8.28 is indeed equal to the Stokes parameter S_1, as we sought.

These results show that the coefficients in the density matrix truly *coincide* with the Stokes parameters. The act of determining these coefficients is tantamount to Stokes polarimetry. In short, the goal of QST is precisely, to determine the Stokes parameters which allow us to reconstruct the complete density matrix.

We now have the perfect recipe for reconstructing density matrices of single-qubit states as long as no errors and perfect experimental conditions are assumed. We merely need to measure $\hat{\rho}$ along the Pauli directions $\{\hat{\sigma}_1, \sigma_2, \sigma_3\}$. We will illustrate this technique in the following section.

8.2 Single-qubit tomography

As stated earlier, state tomography is aimed at estimating the density matrix of an unknown ensemble of particles via a series of carefully planned measurements. In absolute sense, this cannot be performed exactly because it requires an infinitely many number of particles and measurements to eliminate the statistical error.

Though it is unrealistic, we will assume no sources of error in this section and march ahead with the thought experiment. This exercise will employ the toolbox of matrices and projections that we have seen in the previous section. We will explore two-qubit tomography later in this chapter, while considering sources of errors and their compensation.

Assuming perfectly exact measurements, single-qubit state tomography requires making sets of three linearly independent measurements. Let us see what this means. From equations (8.23), (8.24), and (8.25), we understand that we require to measure *six* probabilities, $P_{|D\rangle}$, $P_{|A\rangle}$, $P_{|L\rangle}$, $P_{|R\rangle}$, $P_{|H\rangle}$ and $P_{|V\rangle}$. There are several ways in which this can be achieved. We describe the approach aligned to what we have actually employed in our experiments.

Suppose we have *two* detectors and *three* different kinds of beam splitters at our disposal. These are shown in figure 8.2. Part a shows a beam splitter that routes the input photon into possible output channels depending on the polarization state $\hat{\rho}$ of the input beam. One is the $|H\rangle$ channel and the other is the $|V\rangle$ channel. Let us call this an HV beam splitter. The photons emerging from the channels can be counted with the help of single-photon counting modules. The normalized counts determine the probabilities $P_{|H\rangle}$ and $P_{|V\rangle}$ from which S_3 (as well as $S_0 = P_{|H\rangle} + P_{|V\rangle}$) can be determined. In terms of the Bloch sphere picture, this helps determine how high or low the state $\hat{\rho}$ is with respect to the $\hat{\sigma}_2$–$\hat{\sigma}_3$ plane. The measurement thus collapses the unknown state onto a plane parallel to $\hat{\sigma}_2$–$\hat{\sigma}_3$. We perform these measurements on an

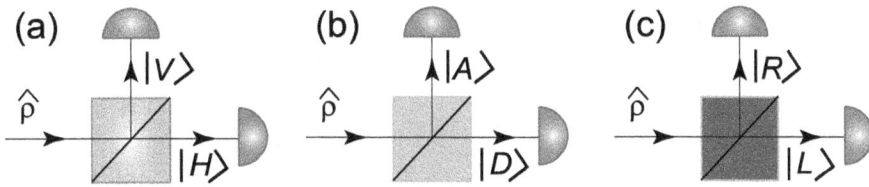

Figure 8.2. Three settings with two detectors gives six different kinds of measurements. Subfigures (a), (b), and (c) are conceptual measurement schemes using three different kinds of beam splitters, the HV, DA and LR beam splitters respectively.

ensemble of identical single photons. No doubt, there should be a large enough number to allow the counts approach the expected values.

We then switch to another beam splitter, say the DA beam splitter which routes the state into $|D\rangle$ and $|A\rangle$ channels. This is shown in figure 8.2(b). These measurements yield $P_{|D\rangle}$ and $P_{|A\rangle}$ allowing us to deduce S_1, collapsing the possible planar solution to a line parallel to the $\hat{\sigma}_1$ axis. Finally, measurement with the LR beam splitter determines S_2 pinpointing a single point (of course with some uncertainty) on the line extracted from the previous experiment.

In summary, before the measurement, the unknown state may be anywhere on or inside the Bloch sphere—we do not know. The first set of measurements always caps the state under investigation to a plane. The second set of measurements isolates the unknown state from the plane to a line. From this line, the third set of measurements finally fetches a point on the Bloch sphere, corresponding to an estimate of the measurand state.

Let us recount the physical resources required for this tomographic procedure. We have three settings corresponding to three different kinds of experimental arrangements (in this case three different kinds of physically distinct beam splitters). We make two measurements within each setting, making our total measurements number to six. From these six outcomes, we deduce the Stokes parameters which allow us to estimate the density matrix.

But hold on, there is a problem. In the laboratory, one may not find these three kinds of rather unusual beam splitters. The HV beam splitter is a common laboratory item but achieving projections in the $\{|D\rangle, |A\rangle\}$ and $\{|L\rangle, |R\rangle\}$ bases is not clear. In order to bypass this, one can follow the measurement scheme outlined in considerable detail in section 5.1.1. This scheme involves placing wave plates in front of the HV beam splitter and orienting them to create the required measurement basis. Two strategies exist. Table 5.1 describes the approach when we put a HWP or a quarter-wave plate (QWP), one by one, whereas table 5.2 delineates the method to simultaneously use both a QWP followed by a HWP, with reorientation of their axes.

This deserves a bit of an elaboration. The basic philosophy goes as follows. Suppose that we need to project the input state along the $\{|D\rangle, |A\rangle\}$ basis. Instead of searching for a beam splitter that can analyze the beam in this fashion, why not rotate the input state so that $|D\rangle$ is rotated to $|H\rangle$ and $|A\rangle$ is rotated to $|V\rangle$ and still keep to the use of HV beam splitter. Thus, the statistics of the detector placed inside the $|H\rangle$ channel firing for this arrangement will be identical to the statistics achieved

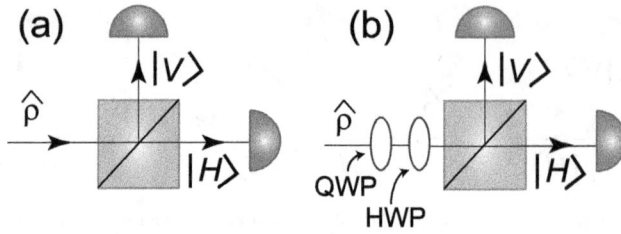

Figure 8.3. The measurement basis can be transformed by adopting an active approach of rotating the input state $\hat{\rho}$: (a) shows how to measure S_3 which is a projection in the $\{|H\rangle, |V\rangle\}$ basis, while (b) shows how the placement of properly oriented QWP and HWP in the path of photons can rotate the state so that it reorients with the rotated measurement basis.

by looking at the detector placed inside the $|D\rangle$ channel of a fictitious DA beam splitter. Therefore, corresponding to the measurements suggested in figures 8.2(b) and (c), we can use a combination of QWP and a HWP oriented at angles q and h. The setup is schematically depicted in figure 8.3. For analysis along the $\{|D\rangle, |A\rangle\}$ basis, we require $q = 45°$, $h = 22.5°$ and for analysis along the $\{|L\rangle, |R\rangle\}$ basis, we use the setting $q = 45°$, $h = 0°$. Furthermore, we can also do the $\{|H\rangle, |V\rangle\}$ analysis with this arrangement by setting $q = 0°$, $h = 0°$. These are just the entries in table 5.2, where the process is described in further detail.

Effectively, the impact of this scheme is to rotate the density matrix (living in the Bloch sphere) so that the desired axes coincide, in a systematic fashion, with the canonical axes of the HV beam splitter.

Recalling experiment Q3, we generated single-qubit states and performed measurements in the aforementioned bases. Hence, we can use the same measurements to estimate the respective density matrices. The procedure is as follows.

Measurements in the $\{|H\rangle, |V\rangle\}$, $\{|D\rangle, |A\rangle\}$, and $\{|L\rangle, |R\rangle\}$ bases let us determine the set of probabilities $P_{|H\rangle}$, $P_{|V\rangle}$, $P_{|D\rangle}$, $P_{|A\rangle}$, $P_{|L\rangle}$, and $P_{|V\rangle}$. Using these probabilities, we calculate the Stokes parameters using the relations given in equation (8.19). Finally, the density matrices are computed using equation (8.15). The results, generated from measurements made in experiment Q3, are presented in table 8.1.

Hence, the single-qubit tomography experiment is straightforward. For an arbitrary one-qubit state, measurements are performed in three canonical bases. Photocounts are used to determine probabilities, which are employed to calculate Stokes parameters. The Stokes parameters finally fetch us the required density matrix. Table 8.1 also shows that the estimated density matrices correlate well with the theoretically predicted ones. We can now generalize this method for two-qubit tomography, and this is where we consider the deleterious impact of imperfections and consider ways to mitigate it.

8.3 Two-qubit tomography

It is reasonably straightforward to generalize the tomography scheme based on the Stokes parameters to measure multi-qubit states. However, it must be kept in mind that there exist important differences between one-photon and two-photon cases.

Table 8.1. Results of single-qubit tomography.

State	Predicted density matrix	Measured density matrix
$\lvert H \rangle$	$\begin{pmatrix} 1 & 0 \\ 0 & 0 \end{pmatrix}$	$\begin{pmatrix} 0.99 & -0.03 - 0.34i \\ -0.03 + 0.34i & 0.01 \end{pmatrix}$
$\lvert V \rangle$	$\begin{pmatrix} 0 & 0 \\ 0 & 1 \end{pmatrix}$	$\begin{pmatrix} 0.06 & 0.10 + 0.24i \\ 0.10 - 0.24i & 0.94 \end{pmatrix}$
$\lvert D \rangle$	$\begin{pmatrix} 0.5 & 0.5 \\ 0.5 & 0.5 \end{pmatrix}$	$\begin{pmatrix} 0.56 & 0.35 - 0.07i \\ 0.35 + 0.07i & 0.44 \end{pmatrix}$
$\lvert A \rangle$	$\begin{pmatrix} 0.5 & -0.5 \\ -0.5 & 0.5 \end{pmatrix}$	$\begin{pmatrix} 0.49 & -0.45 - 0.04i \\ -0.45 + 0.04i & 0.51 \end{pmatrix}$
$\lvert L \rangle$	$\begin{pmatrix} 0.5 & -0.5i \\ 0.5i & 0.5 \end{pmatrix}$	$\begin{pmatrix} 0.54 & 0.04 - 0.48i \\ 0.04 + 0.48i & 0.46 \end{pmatrix}$
$\lvert R \rangle$	$\begin{pmatrix} 0.5 & 0.5i \\ -0.5i & 0.5 \end{pmatrix}$	$\begin{pmatrix} 0.44 & 0.02 + 0.44i \\ 0.02 - 0.44i & 0.56 \end{pmatrix}$

For instance, one-photon beams demonstrate polarization properties similar to those of classical optical beams, and hence can be described in a purely classical manner [19] if differences in photodetection statistics are ignored. For two-photon or bipartite systems, however, nonclassical correlations may also exist between the two or more beams. This is due to the unusual, purely quantum property of quantum entanglement.

In this section, we are interested in two-photon states. These states live in a four-dimensional space and are represented by a 4×1 column vector (if pure). For example, an arbitrary two-photon pure state can be written as

$$\lvert \psi \rangle = a_0 \lvert HH \rangle + a_1 \lvert HV \rangle + a_2 \lvert VH \rangle + a_3 \lvert VV \rangle = \begin{pmatrix} a_0 \\ a_1 \\ a_2 \\ a_3 \end{pmatrix} \qquad (8.30)$$

with complex numbers a_i and $\sum_{i=0}^{3} \lvert a_i \rvert^2 = 1$. The generalized mixed state can be written as the density matrix

$$\hat{\rho} = \begin{pmatrix} A_1 & B_1 e^{i\phi_1} & B_2 e^{i\phi_2} & B_3 e^{i\phi_3} \\ B_1 e^{-i\phi_1} & A_2 & B_4 e^{i\phi_4} & B_5 e^{i\phi_5} \\ B_2 e^{-i\phi_2} & B_4 e^{-i\phi_4} & A_3 & B_6 e^{i\phi_6} \\ B_3 e^{-i\phi_3} & B_5 e^{-i\phi_5} & B_6 e^{-i\phi_6} & A_4 \end{pmatrix}, \qquad (8.31)$$

which makes $\hat{\rho}$ a semi-definite positive Hermitian and gives a trace equal to one ($A_1 + A_2 + A_3 + A_4 = 1$). Hence, we need $16 - 1 = 15$ parameters to completely determine this matrix. These parameters are ($A_1, \ldots, A_3, B_1, \ldots, B_6, \phi_1, \ldots, \phi_6$). For this two-qubit state, the equation analogous to equation (8.15) is

$$\hat{\rho} = \frac{1}{4} \sum_{i,j=0}^{3} S_{ij} \hat{\sigma}_i \otimes \hat{\sigma}_j,$$ (8.32)

and the S coefficients are generally given as

$$S_{ij} = S_i \otimes S_j = (P_{|\psi_i\rangle} \pm P_{|\psi_i\perp\rangle}) \otimes (P_{|\psi_j\rangle} \pm P_{|\psi_j\perp\rangle}).$$ (8.33)

In this notation, $P_{|\psi_i\rangle}$ is the probability of projecting a single qubit state to the output channel designated as $|\psi_i\rangle$ in a beam splitter experiment that differentiates between $|\psi_i\rangle$ and its orthogonal state $|\psi_i \perp \rangle$. For normalization, it is required that $S_{00} = 1$. Hence, 15 real parameters are required to identify any point in the two-qubit four-dimensional Hilbert space. These are in fact two-qubit versions of the Stokes parameters for single-qubit states. However, unlike the Bloch sphere for one-qubit states, we do not have an accessible graphical picture of this two-qubit space. Yet, the Stokes parameters and the measurement probabilities are still related in the two-qubit space [2, 20].

The two-qubit S coefficients can also be spelled out showing the meaning of the notation for the tensor product $\otimes$ used in equation (8.33),

$$
\begin{aligned}
S_{00} &= (P_{|H\rangle} + P_{|V\rangle}) \otimes (P_{|H\rangle} + P_{|V\rangle}) = P_{|HH\rangle} + P_{|HV\rangle} + P_{|VH\rangle} + P_{|VV\rangle} \\
S_{01} &= (P_{|H\rangle} + P_{|V\rangle}) \otimes (P_{|D\rangle} - P_{|A\rangle}) = P_{|HD\rangle} - P_{|HA\rangle} + P_{|VD\rangle} - P_{|VA\rangle} \\
S_{02} &= (P_{|H\rangle} + P_{|V\rangle}) \otimes (P_{|L\rangle} - P_{|R\rangle}) = P_{|HL\rangle} - P_{|HR\rangle} + P_{|VL\rangle} - P_{|VR\rangle} \\
S_{03} &= (P_{|H\rangle} + P_{|V\rangle}) \otimes (P_{|H\rangle} - P_{|V\rangle}) = P_{|HH\rangle} - P_{|HV\rangle} + P_{|VH\rangle} - P_{|VV\rangle} \\
S_{10} &= (P_{|D\rangle} - P_{|A\rangle}) \otimes (P_{|H\rangle} + P_{|V\rangle}) = P_{|DH\rangle} + P_{|DV\rangle} - P_{|AH\rangle} - P_{|AV\rangle} \\
S_{11} &= (P_{|D\rangle} - P_{|A\rangle}) \otimes (P_{|D\rangle} - P_{|A\rangle}) = P_{|DD\rangle} - P_{|DA\rangle} - P_{|AD\rangle} + P_{|AA\rangle} \\
S_{12} &= (P_{|D\rangle} - P_{|A\rangle}) \otimes (P_{|L\rangle} - P_{|R\rangle}) = P_{|DL\rangle} - P_{|DR\rangle} - P_{|AL\rangle} + P_{|AR\rangle} \\
S_{13} &= (P_{|D\rangle} - P_{|A\rangle}) \otimes (P_{|H\rangle} - P_{|V\rangle}) = P_{|DH\rangle} - P_{|DV\rangle} - P_{|AH\rangle} + P_{|AV\rangle} \\
S_{20} &= (P_{|L\rangle} - P_{|R\rangle}) \otimes (P_{|H\rangle} + P_{|V\rangle}) = P_{|LH\rangle} + P_{|LV\rangle} - P_{|RH\rangle} - P_{|RV\rangle} \\
S_{21} &= (P_{|L\rangle} - P_{|R\rangle}) \otimes (P_{|D\rangle} - P_{|A\rangle}) = P_{|LD\rangle} - P_{|LA\rangle} - P_{|RD\rangle} + P_{|RA\rangle} \\
S_{22} &= (P_{|L\rangle} - P_{|R\rangle}) \otimes (P_{|L\rangle} - P_{|R\rangle}) = P_{|LL\rangle} - P_{|LR\rangle} - P_{|RL\rangle} + P_{|RR\rangle} \\
S_{23} &= (P_{|L\rangle} - P_{|R\rangle}) \otimes (P_{|H\rangle} - P_{|V\rangle}) = P_{|LH\rangle} - P_{|LV\rangle} - P_{|RH\rangle} + P_{|RV\rangle} \\
S_{30} &= (P_{|H\rangle} - P_{|V\rangle}) \otimes (P_{|H\rangle} + P_{|V\rangle}) = P_{|HH\rangle} + P_{|HV\rangle} - P_{|VH\rangle} - P_{|VV\rangle} \\
S_{31} &= (P_{|H\rangle} - P_{|V\rangle}) \otimes (P_{|D\rangle} - P_{|A\rangle}) = P_{|HD\rangle} - P_{|HA\rangle} - P_{|VD\rangle} + P_{|VA\rangle} \\
S_{32} &= (P_{|H\rangle} - P_{|V\rangle}) \otimes (P_{|L\rangle} - P_{|R\rangle}) = P_{|HL\rangle} - P_{|HR\rangle} - P_{|VL\rangle} + P_{|VR\rangle} \\
S_{33} &= (P_{|H\rangle} - P_{|V\rangle}) \otimes (P_{|H\rangle} - P_{|V\rangle}) = P_{|HH\rangle} - P_{|HV\rangle} - P_{|VH\rangle} + P_{|VV\rangle}.
\end{aligned}
$$ (8.34)

In these 16 formulations, the term $P_{|HD\rangle}$, for example, refers to the joint probability of detecting the first photon in the channel $|H\rangle$ when it is subject to an HV beam splitter and the second photon is detected in the channel $|D\rangle$ when this second photon is analyzed by an 'effective' DA beamsplitter. All other definitions are analogous.

Now, in order to estimate the density matrix, we need to determine all the S coefficients. We can also upgrade the discussion on resource requirements for the single qubit to the two-qubit case. Suppose we use two detectors for each qubit, a total of four. For each qubit, we can use three settings corresponding to the projections along the three bases: $\{|D\rangle, |A\rangle\}, \{|L\rangle, |R\rangle\}$ and $\{|H\rangle, |V\rangle\}$. This means that $3 \times 3 = 3^2 = 9$ settings are required. In each setting, we measure four ($2^2 = 4$) probabilities corresponding to the output channels $|HH\rangle, |HV\rangle, |VH\rangle, |VV\rangle$. This shows that $9 \times 4 = 36$ measurements are sufficient to provide information about the state $\hat{\rho}$. But there are only 15 independent parameters, so clearly we have more measurements than are actually needed. We have an overdetermined system.

The density matrix can be used as a tool to characterize any state, determine the purity, degree of mixedness or quantify the amount of entanglement. For instance, *fidelity* is a quantity defined to measure the state overlap or 'likeness' of two states. For states represented by density matrices $\hat{\rho}_1$ and $\hat{\rho}_2$, fidelity F is generally given by [17]

$$F(\hat{\rho}_1, \hat{\rho}_2) = (Tr[\sqrt{\sqrt{\hat{\rho}_1} \hat{\rho}_2 \sqrt{\hat{\rho}_1}}\,])^2. \tag{8.35}$$

Furthermore, if one of the two states being compared is pure, then the fidelity expression simplifies to

$$F(\hat{\rho}_1, \hat{\rho}_2) = Tr(\hat{\rho}_1 \hat{\rho}_2). \tag{8.36}$$

Hence, if we want to check how well a particular entangled state is generated, we can do QST of the generated state and check its fidelity against the theoretical prediction.

Two other figures of merit are called *concurrence* and *tangle*. They closely related measures are used to quantify the entanglement of a system [8, 21, 22]. For a two-qubit system such as ours, concurrence C is defined as

$$C = \text{Max}\left\{0, \sqrt{\lambda_1} - \sqrt{\lambda_2} - \sqrt{\lambda_3} - \sqrt{\lambda_4}\right\} \tag{8.37}$$

where $\lambda_1 > \lambda_2 > \lambda_3 > \lambda_4$ are the eigenvalues of the matrix $\hat{\rho}\hat{Z}\hat{\rho}^T\hat{Z}$. Superscript T denotes the transpose function and $\hat{Z}$ is called a 'spin flip matrix', which is defined as

$$\hat{Z} \equiv \begin{pmatrix} 0 & 0 & 0 & -1 \\ 0 & 0 & 1 & 0 \\ 0 & 1 & 0 & 0 \\ -1 & 0 & 0 & 0 \end{pmatrix}. \tag{8.38}$$

Tangle T is simply obtained from concurrence using the relation $T = C^2$. Both concurrence and tangle range from 0 for maximally mixed states to 1 for maximally entangled states.

8.4 Nonideal measurements and compensation of errors

The technique described above empowers us to perfectly reconstruct the density matrix, but only if we have infinitely many ideal measurements. In real experiments, we cannot measure ideal probabilities and measurement is also never perfect.

Moreover, imperfect wave plate orientations also result in measurements that are in bases which may be slightly different from the intended ones. On the other hand, a density matrix corresponding to any physical state must be positive semi-definite. Coupled with normalization and Hermiticity, this implies that all the eigenvalues of the density matrix must lie between 0 and 1 (inclusive), and that their sum must be 1. This, in turn, implies that $0 < Tr(\hat{\rho}) < 1$. It is well known that these conditions are not achieved in all experimental tomographic measurements [8, 23].

The errors propagating into the density matrix can be categorized into three types [8]. First, we can have errors in the measurement basis. These errors can be tackled by using an increasingly accurate measurement apparatus. Second, we have errors resulting from the counting statistics. Such errors can be reduced by taking a larger number of measurements, i.e., recording photocounts for a longer time. Third, we may have errors due to lack of experimental stability. These errors originate from variability of the generated state or the instability of the measurement apparatus. The variability in the detected intensity of the state (i.e., the rate of photons produced) is called drift. It can be somewhat compensated by using four detectors (instead of two) for the two-photon case [23]. The idea is that since each bipartite member of an ensemble is measured in a complete basis, we do not need to assume a constant ensemble size for each measurement. For example, using a four-detector system with $\{|H\rangle, |V\rangle\}$ basis, we can obtain the photocounts N_{HH}, N_{HV}, N_{VH}, and N_{VV} for a finite time period. From these counts, total photocounts and probabilities can be determined confidently. If there is a drift in state intensity during a change in measurement basis, it will not harm the tomography results significantly because the total ensemble size is determined separately for each measurement.

As discussed in chapter 4, there are accidental coincidence counts that are contributed by background light. Then there are the detector's dark counts. These counts can all be subtracted from the total measured counts.

Even after taking care of all of these errors, it is still possible for the state estimation procedure to fetch us an illegal density matrix (e.g., a single-qubit state with radius greater than 1 in the Bloch space). This problem is taken care of by finding instead the legal state that may most likely have returned the counts that have been measured [2, 3]. This is the essence of the maximum-likelihood technique, which is utilized in the QST experiments briefly explained in the next section. Maximum-likelihood estimation finds the parameters that may have made the outcomes most likely, provided that a certain model is assumed. Our unknown parameters are the elements of the density matrix $\hat{\rho}$, which are given by $\{A_1, A_2, A_3, A_4, B_1, B_2, B_3, B_4, B_5, B_6, \phi_1, \phi_2, \phi_3, \phi_4, \phi_5, \phi_6\}$ with $\sum_{i=0}^{4} A_i = 1$.

8.5 Maximum-likelihood estimation

The maximum-likelihood technique provides an effective way to accommodate for imperfect measurements [8]. This technique resolves the issue of illegal density matrices by finding the state that is most likely to have resulted in the measured photocounts [2, 3]. Determining this legitimate state analytically is an extremely nontrivial task, and therefore we resort to numerical optimization. Maximum-likelihood estimation requires three major elements: expression of a general density matrix in terms of legitimate parameters, a likelihood function that can be maximized, and a numerical technique for performing this optimization. The numerical technique must perform this maximization over the space of density matrix parameters.

Here, a legitimate state implies one having a non-negative definite Hermitian density matrix with a unit trace. The density matrix is expressed in the form of an accompanying matrix $\hat{W}$ such that

$$\hat{\rho}_p = \frac{\hat{W}^\dagger \hat{W}}{Tr\left\{\hat{W}^\dagger \hat{W}\right\}} \tag{8.39}$$

fulfills the aforementioned requirements for valid density matrices for physical systems [2]. The subscript p in $\hat{\rho}_p$ signifies a physically legitimate density matrix. Dealing with the two-qubit system, it is a 4×4 density matrix with 15 independent real parameters. It is convenient to choose a tri-diagonal form for $\hat{W}$ [2] given by

$$\hat{W} = \begin{pmatrix} w_1 & 0 & 0 & 0 \\ w_5 + iw_6 & w_2 & 0 & 0 \\ w_7 + iw_8 & w_9 + iw_{10} & w_3 & 0 \\ w_{11} + iw_{12} & w_{13} + iw_{14} & w_{15} + iw_{16} & w_4 \end{pmatrix}. \tag{8.40}$$

Consider the measurement data consisting of a set of 36 coincidence counts n_i whose expected values are given by $\bar{n}_i = N\langle\psi_i|\hat{\rho}_p|\psi_i\rangle$ (for $i = 1, 2, \ldots, 36$), where N is a normalization parameter equal to the size of the ensemble per measurement. Then, assuming a Gaussian probability distribution of the coincidence counts, the optimization problem can be stated as being equivalent to finding the minimum of the function [2]

$$\mathscr{L} = \sum_{i=1}^{36} \frac{\left[N\langle\psi_i|\hat{\rho}_p|\psi_i\rangle - n_i\right]^2}{2N\langle\psi_i|\hat{\rho}_p|\psi_i\rangle}. \tag{8.41}$$

where n_i is the result of i'th measurement. The function $\mathscr{L}$ is called the log-likelihood function. The final step in the method is an optimization routine. We use the MATLAB toolbox developed by the Kwiat group[2][8], and will now discuss the experiment.

8.6 The experiment

We use a four-detector system for the two qubits. The experimental setup is an extension to that of Experiments NL2–NL3. Only two QWPs are added in the paths of the two downconverted beams. A schematic diagram of the setup is shown in

---

[2] http://research.physics.illinois.edu/QI/Photonics/tomography/

figure 8.4, whereas a photograph is shown in figure 8.5. This is the extension of the single-qubit case illustrated in figure 8.3(b) to two qubits.

As shown in experiment Q3, we can perform an arbitrary polarization measurement and its orthogonal complement using a QWP, a HWP, and a PBS. During the tomography experiment, we only change the orientation of the wave plates using motorized mounts to realize measurements in different polarization bases, whereas

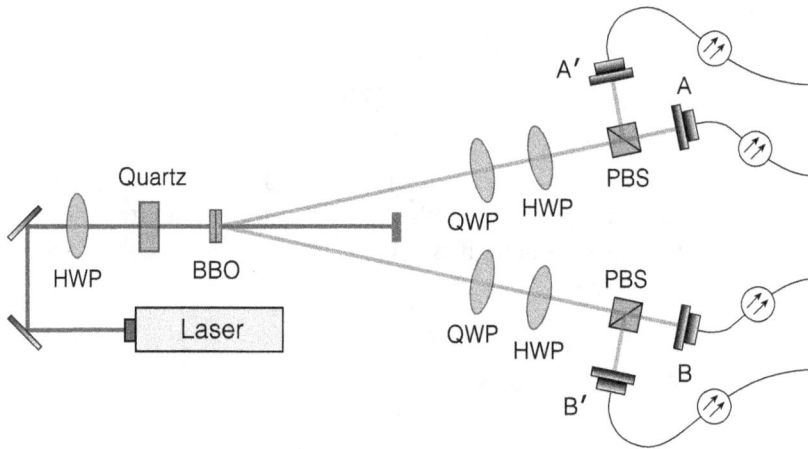

Figure 8.4. Schematic diagram of the two-qubit tomography experiment. In each path of the downconverted photon pair, a QWP, a HWP, a polarizing beam splitter (PBS), and two detectors are placed. The pump beam HWP and quartz plate (Q) are used to generate the required input state.

Figure 8.5. Photograph of the two-qubit tomography setup, excluding the quartz plate.

the input states are modified by changing the orientation of the pump beam HWP and rotation of the quartz plate with respect to the vertical axis.

Consider the general state (6.1) with parameters A, B, and φ. The coefficients A and B are controlled by the orientation of the HWP. The phase φ can be effectively modified by changing the rotation of the quartz plate, which is basically a birefringent plate that adds a tunable phase factor to the $|VV\rangle$ term. We have already described the use of this phase-correcting element in section 6.3.3. In this way, we can generate polarization states that are arbitrary linear combinations of the $|HH\rangle$ and $|VV\rangle$ states. For the $|HH\rangle$ and $|VV\rangle$ states, the quartz plate orientation does not matter. However, to synthesize the Bell state, the quartz plate is needed to eliminate the effective phase between the $|HH\rangle$ and $|VV\rangle$ terms.

For the tomography of each generated state, we performed measurements in the nine settings, which along with the respective wave plate orientations are tabulated in table 8.2. Here we use the notations h_A, q_A, h_B, and q_B for the HWP and QWP orientations for Alice (A) and Bob (B). All orientations are measured with respect to the laboratory's horizontal axis.

We perform tomography for three kinds of generated states $|HH\rangle$, $|VV\rangle$, and $(|HH\rangle + |VV\rangle)/\sqrt{2}$, using Kwiat's maximum-likelihood estimation toolbox, compare the estimated density matrices with the theoretically predicted ones and compute the fidelity for each generated state (table 8.6). The datasets for the three states are presented in tables 8.3, 8.4, and 8.5, respectively. Finally, the estimated density matrices are also plotted in figure 8.6.

As discussed earlier, we can determine the purity of a state from its density matrix. For example, let us take the estimated density matrix for the Bell state $(|HH\rangle + |VV\rangle)/\sqrt{2}$. Its purity $Tr(\hat{\rho}^2)$ is calculated to be 0.68. In contrast, if we take the predicted matrix which we also calculated in equation (3.36) in chapter 3, then its purity comes out to be 1. For the generated Bell state, we also calculate concurrence (=0.61) and tangle (=0.37), which unsurprisingly are not equal to the ideal value

Table 8.2. Measurement bases for two-qubit tomography. h_A (h_B) and q_A (q_B) represent the HWP and QWP orientations for Alice (Bob).

Measurement basis	h_A	q_A	h_B	q_B
$|HH\rangle$	0°	0°	0°	0°
$|HD\rangle$	0°	0°	22.5°	45°
$|HL\rangle$	0°	0°	0°	45°
$|DH\rangle$	22.5°	45°	0°	0°
$|DD\rangle$	22.5°	45°	22.5°	45°
$|DL\rangle$	22.5°	45°	0°	45°
$|LH\rangle$	0°	45°	0°	0°
$|LD\rangle$	0°	45°	22.5°	45°
$|LL\rangle$	0°	45°	0°	45°

Table 8.3. Measurements for the two-qubit tomography of the $|HH\rangle$ state. The numbers represent average measured coincidence counts per second. Accidental coincidence counts have been subtracted.

Measurement basis	AB	AB'	$A'B$	$A'B'$
$|HH\rangle$	1691	41	1	0
$|HD\rangle$	695	1124	0	7
$|HL\rangle$	1414	441	0	6
$|DH\rangle$	1004	20	791	0
$|DD\rangle$	405	620	339	496
$|DL\rangle$	857	196	302	208
$|LH\rangle$	852	5	203	0
$|LD\rangle$	322	165	72	449
$|LL\rangle$	666	54	90	99

Table 8.4. Measurements for the two-qubit tomography of the $|VV\rangle$ state. The numbers represent average measured coincidence counts per second. Accidental coincidence counts have been subtracted.

Measurement basis	AB	AB'	$A'B$	$A'B'$
$|HH\rangle$	44	3	103	2031
$|HD\rangle$	24	10	958	808
$|HL\rangle$	47	2	1016	1583
$|DH\rangle$	13	719	73	1173
$|DD\rangle$	311	278	509	458
$|DL\rangle$	155	557	412	1029
$|LH\rangle$	52	1287	80	1365
$|LD\rangle$	508	445	847	662
$|LL\rangle$	146	961	874	1003

($=1$). The take-home message is that we desired to generate a pure, entangled state but due to nonideal experimental conditions we obtained an impure, mixed state.

Analyzing errors in the experimentally reconstructed density matrices is not a straightforward task. We do not perform error analysis here but briefly mention two methods. The traditional method analytically accounts for the errors in the measurements due to each identified source of error. The measurement errors are then propagated through calculations of derived quantities [2]. In this way, errors in the density matrices and its derived quantities, due to counting statistics and

Table 8.5. Measurements for the two-qubit tomography of the Bell state. The numbers represent average measured coincidence counts per second. Accidental coincidence counts have been subtracted.

Measurement basis	AB	AB'	$A'B$	$A'B'$
$\lvert HH \rangle$	836	50	18	896
$\lvert HD \rangle$	433	615	297	437
$\lvert HL \rangle$	520	729	50	652
$\lvert DH \rangle$	566	197	371	459
$\lvert DD \rangle$	715	145	91	766
$\lvert DL \rangle$	354	596	371	242
$\lvert LH \rangle$	215	284	630	111
$\lvert LD \rangle$	332	621	324	360
$\lvert LL \rangle$	84	733	641	287

Table 8.6. Results of two-qubit QST. For the estimated density matrices, only the real components of the matrices are shown.

State	Predicted density matrix	Re{Measured density matrix}	Fidelity
$\lvert HH \rangle$	$\begin{pmatrix} 1 & 0 & 0 & 0 \\ 0 & 0 & 0 & 0 \\ 0 & 0 & 0 & 0 \\ 0 & 0 & 0 & 0 \end{pmatrix}$	$\begin{pmatrix} 0.92 & -0.09 & 0.05 & -0.04 \\ -0.09 & 0.05 & 0.02 & 0.02 \\ 0.05 & 0.02 & 0.02 & 0.01 \\ -0.04 & 0.02 & 0.01 & 0.01 \end{pmatrix}$	0.92
$\lvert VV \rangle$	$\begin{pmatrix} 0 & 0 & 0 & 0 \\ 0 & 0 & 0 & 0 \\ 0 & 0 & 0 & 0 \\ 0 & 0 & 0 & 1 \end{pmatrix}$	$\begin{pmatrix} 0.01 & 0.00 & -0.01 & 0.06 \\ 0.00 & 0.02 & 0.00 & -0.10 \\ -0.01 & 0.00 & 0.05 & 0.04 \\ 0.06 & -0.10 & 0.04 & 0.92 \end{pmatrix}$	0.92
$\frac{\lvert HH \rangle + \lvert VV \rangle}{\sqrt{2}}$	$\begin{pmatrix} 0.5 & 0 & 0 & 0.5 \\ 0 & 0 & 0 & 0 \\ 0 & 0 & 0 & 0 \\ 0.5 & 0 & 0 & 0.5 \end{pmatrix}$	$\begin{pmatrix} 0.38 & -0.01 & -0.05 & 0.30 \\ -0.01 & 0.05 & 0.00 & 0.06 \\ -0.05 & 0.00 & 0.02 & -0.05 \\ 0.30 & 0.06 & -0.05 & 0.55 \end{pmatrix}$	0.77

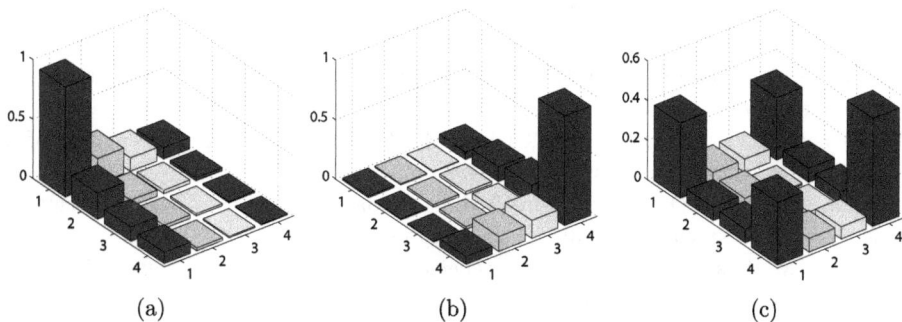

(a)   (b)   (c)

Figure 8.6. Experimentally estimated density matrices for (a) $\lvert HH \rangle$, (b) $\lvert VV \rangle$, and (c) $(\lvert HH \rangle + \lvert VV \rangle)/\sqrt{2}$ states. The plots are made using the absolute value of the respective density matrix components.

imperfect wave plates, are analyzed. Another method uses the Monte Carlo technique [8]. This technique revolves around using numerically simulated data to obtain additional density matrices, which are then exploited to compute errors in the quantities pertaining to the estimated density matrix.

References

[1] Stokes G G 1851 On the composition and resolution of streams of polarized light from different sources *Trans. Camb. Phil. Soc.* **9** 399

[2] James D F V, Kwiat P G, Munro W J and White A G 2005 On the measurement of qubits *Asymptotic Theory of Quantum Statistical Inference: Selected Papers* (Singapore: World Scientific) pp 509–38

[3] Hradil Z 1997 Quantum-state estimation *Phys. Rev.* A **55** R1561

[4] Tan S M 1997 An inverse problem approach to optical homodyne tomography *J. Mod. Opt.* **44** 2233–59

[5] Banaszek K, D'ariano G M, Paris M G A and Sacchi M F 1999 Maximum-likelihood estimation of the density matrix *Phys. Rev.* A **61** 010304

[6] Hradil Z, Summhammer J, Badurek G and Rauch H 2000 Reconstruction of the spin state *Phys. Rev.* A **62** 014101

[7] Řeháček J, Hradil Z and Ježek M 2001 Iterative algorithm for reconstruction of entangled states *Phys. Rev.* A **63** 040303

[8] Altepeter J B, Jeffrey E R and Kwiat P G 2005 Photonic state tomography *Adv. Atom. Mol. Opt. Phy.* **52** 105–59

[9] Ashburn J R, Cline R A, van der Burgt P J M, Westerveld W B and Risley J S 1990 Experimentally determined density matrices for h (n = 3) formed in h+–he collisions from 20 to 100 kev *Phys. Rev.* A **41** 2407

[10] Smithey D T, Beck M, Raymer M G and Faridani A 1993 Measurement of the Wigner distribution and the density matrix of a light mode using optical homodyne tomography: application to squeezed states and the vacuum *Phys. Rev. Lett.* **70** 1244

[11] Dunn T J, Walmsley I A and Mukamel S 1995 Experimental determination of the quantum-mechanical state of a molecular vibrational mode using fluorescence tomography *Phys. Rev. Lett.* **74** 884

[12] Leibfried D, Meekhof D M, King B E, Monroe C H, Itano W M and Wineland D J 1996 Experimental determination of the motional quantum state of a trapped atom *Phys. Rev. Lett.* **77** 4281

[13] Kurtsiefer Ch, Pfau T and Mlynek J 1997 Measurement of the Wigner function of an ensemble of helium atoms *Nature* **386** 150

[14] Klose G, Smith G and Jessen P S 2001 Measuring the quantum state of a large angular momentum *Phys. Rev. Lett.* **86** 4721

[15] Chuang I L, Gershenfeld N and Kubinec M 1998 Experimental implementation of fast quantum searching *Phys. Rev. Lett.* **80** 3408

[16] White A G, James D F V, Eberhard P H and Kwiat P G 1999 Nonmaximally entangled states: production, characterization, and utilization *Phys. Rev. Lett.* **83** 3103

[17] Nielsen M A and Chuang I 2010 *Quantum Computation and Quantum Information* (Cambridge: Cambridge University Press)

[18] Hecht E 1998 *Optics* (Reading, MA: Addison-Wesley)

[19] Mandel L and Wolf E 1995 *Optical Coherence and Quantum Optics* (Cambridge: Cambridge University Press)

[20] Abouraddy A F, Sergienko A V, Saleh B E A and Teich M C 2002 Quantum entanglement and the two-photon stokes parameters *Opt. Commun.* **201** 93–8

[21] Wootters W K 1998 Entanglement of formation of an arbitrary state of two qubits *Phys. Rev. Lett.* **80** 2245

[22] Coffman V, Kundu J and Wootters W K 2000 Distributed entanglement *Phys. Rev.* A **61** 052306

[23] Altepeter J B, James D F V and Kwiat P G 2004 *4 Qubit quantum state tomography Quantum State Estimation* (Berlin: Springer) pp 113–45

Chapter 9

Conclusion

Nowadays, many experiments are possible in the instructional physics laboratory using single photons. These experiments are modern, cost-effective versions of some of the most groundbreaking experiments that shaped our understanding of quantum mechanics in the previous century. Some of these experiments include demonstrating the existence of single photons [1], tests of Bell inequalities [2–4], single-photon interference [5], quantum eraser [5], and quantum state tomography [6].

As narrated in this book, which might be used as a textbook for the laboratory, we have developed all the aforementioned experiments in our laboratory using a modular approach. In other words, subsequent experiments build on the earlier ones. This approach makes troubleshooting easier and renders confidence to the experimenter because the simpler experiments are set up first.

We would like to comment on the relative difficulty of setting up the various experiments. As far as alignment is concerned, experiment Q1 requires two detectors and is the easiest. Experiment Q4 can be directly built on it with addition of few optical elements. Experiment NL1 also follows from experiment Q1 but the entangled state generation makes the experiment a bit difficult. Next in terms of complexity level are experiments Q2 and Q3, both of which require three detectors. However, once experiment Q2 is set up, experiment Q3 needs very little effort. The next difficult experiments include those involving four detectors and entanglement—experiments NL2, NL3, Q+NL, and QST. We found experiment Q5 and Q+NL, incorporating single-photon interference and quantum erasure, to be the most challenging. Setting up the interferometer and changing the interferometer path difference with the optimal finesse required skill and experience.

It goes without saying that many versions of these experiments have been developed and performed earlier in a number of colleges and universities, most of them in the United States [7–9]. We found the earlier discussions of these experiments to be very helpful in setting up our laboratory. To the best of our knowledge,

doi:10.1088/978-0-7503-6315-0ch9

ours is the first such laboratory in Pakistan. As technology continues to advance, it is expected that more institutions will be able to set up their versions of quantum mechanical laboratories very soon. We hope that they will find this book a valuable resource.

We would like to conclude by mentioning a few of the wide-ranging benefits that these experiments promise. They offer effective pedagogical tools to complement quantum mechanics courses. These experiments can motivate students to think about the fundamental aspects of quantum physics. This can also prove very beneficial for students who would like to investigate the foundations of quantum mechanics—an important area of research these days. Above all, these experiments can teach students a functional framework to think about and do research in quantum optics, quantum computing, quantum communication, and quantum information.

References

[1] Thorn J J, Neel M S, Donato V W, Bergreen G S, Davies R E and Beck M 2004 Observing the quantum behavior of light in an undergraduate laboratory *Am. J. Phys.* **72** 1210–9

[2] Dehlinger D and Mitchell M W 2002 Entangled photons, nonlocality, and Bell inequalities in the undergraduate laboratory *Am. J. Phys.* **70** 903–10

[3] Carlson J A, Olmstead M D and Beck M 2006 Quantum mysteries tested: an experiment implementing Hardy's test of local realism *Am. J. Phys.* **74** 180–6

[4] Brody J and Selton C 2018 Quantum entanglement with Freedman's inequality *Am. J. Phys.* **86** 412–6

[5] Beck M 2012 *Quantum Mechanics: Theory and Experiment* (Oxford: Oxford University Press)

[6] Altepeter J B, Jeffrey E R and Kwiat P G 2005 Photonic state tomography *Adv. Atom. Mol. Opt. Phy.* **52** 105–59

[7] Beck M and Galvez E J 2007 Quantum optics in the undergraduate teaching laboratory *Conference on Coherence and Quantum Optics* (Washington, DC: Optical Society of America) p CSuA4

[8] Galvez E J 2019 Quantum optics laboratories for teaching quantum physics *Education and Training in Optics and Photonics* (Washington, DC: Optical Society of America) Paper 11 143_123

[9] Lukishova S G 2017 Quantum optics and nano-optics teaching laboratory for the under-graduate curriculum: teaching quantum mechanics and nano-physics with photon counting instrumentation In *Education and Training in Optics and Photonics* (Washington, DC: Optical Society of America) p 104522I

IOP Publishing

Quantum Mechanics in the Single-Photon Laboratory
(Second Edition)

Muhammad Sabieh Anwar, Faizan-e-Ilahi, Syed Bilal Hyder and Muhammad Hamza Waseem

Appendix A

FPGA—introduction and programming

A.1 Introduction

Establishing a single-photon quantum physics laboratory from scratch requires a combined effort of physicists, engineers, and computer scientists. However, putting together such a multidisciplinary team is a grueling task, and we are often stuck with a team comprising only physicists. This is where our road of research starts getting bumpy.

Developing a high-speed coincidence counting unit (CCU) on a field programmable gate array (FPGA) [1–3], for example, requires a good command over a hardware description language (HDL). The working principle of FPGAs is based on digital logic, and digital logic designing is a skill that is considered a specialty of electrical engineers. Although there are books available to learn about FPGAs and their programming [4–6], this appendix provides an introduction to FPGA, aimed to guide the physicist who is perhaps encountering FPGAs for the first time.

A.2 Digital logic design

Boolean logic can be thought of as a language whereby we can communicate only with a 'yes' or a 'no.' In a digital implementation of Boolean logic, a 'no' can be communicated with nearly zero voltage, which is numerically represented as '0', and a 'yes' can be communicated by a relatively higher voltage which is numerically represented as a '1', and to communicate something more meaningful we use very long combinations of these. Digital logic design refers to the use of transistors (semiconductor-based devices that switch or amplify electrical signals) to make circuits that can perform rather complex computational tasks. The design starts with using transistors to make the most basic units of digital design, which are called logic gates. Examples include AND and NOT gates. A NOT gate is a circuit that takes in an input logic, reverses it, and outputs it. A '0' becomes a '1', and vice versa.

On the other hand, an AND gate is a circuit that takes in two digital logic inputs and one output. The output will be '1' if and only if both the inputs are '1'.

When digital electronics was still in its infancy, discrete logic chips (e.g., AND, OR, NOT, NAND, etc.) were used to build entire logic circuits. It is still possible, in theory at least, to make any digital circuit using just these chips. However, we prefer not to resort to that technique any longer because of its slow speed, high cost, and implementation complexity. The concept behind logic design has not changed, but now we have much more efficient methods to implement these circuits. To compare these new methods, we would like to mention application-specific integrated circuits (ASIC), microprocessors, and FPGAs.

A.2.1 ASICs

This circuit implementation method takes the lead over many other methods when it comes to speed. ASICs are extremely fast and work exceptionally well for the task they are specifically designed for. However, there is a price to pay for this speed, i.e., these circuits can be programmed only once. One small mistake or a need for change in the implementation of the circuit will render the entire batch of old circuits useless.

A.2.2 Microprocessors

Another commonplace of implementing logic on a chip is through the use of microprocessors. Microprocessors contain a few inflexible digital logic circuits that can perform some simple computational tasks such as arithmetic problems. The task to be performed by a microprocessor gets broken down into smaller tasks that the available fixed circuits can do. These circuits are then used to accomplish the required task. This implementation gets the job done, but to no one's surprise using hardware this way is brutally inefficient.

A.2.3 FPGAs

At this point, the FPGAs come into play. These chips are made with rows of gates stacked over one another, but the gates are left unconnected. Based on the design needs, the gates are connected through wiring channels, giving us circuitry that can get the job done.

A.3 Building blocks of an FPGA

The design of an FPGA can be broken down into three significant components. These are logic blocks, routing channels, and I/O pads. The conceptual scheme of these components is shown in figure A.1.

A.3.1 Logic blocks

A 2-to-1 multiplexer is a simple digital logic circuit made using two AND gates, one NOT gate, and one OR gate. It has a total of three inputs, one of which is a special input known as the selection input.

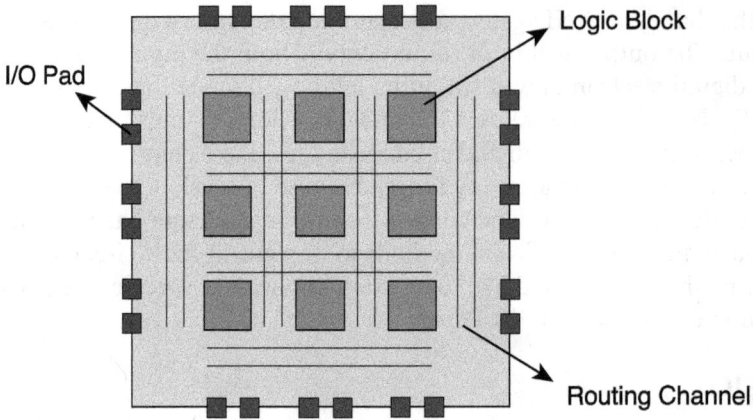

Figure A.1. Illustration for FPGA's overall design.

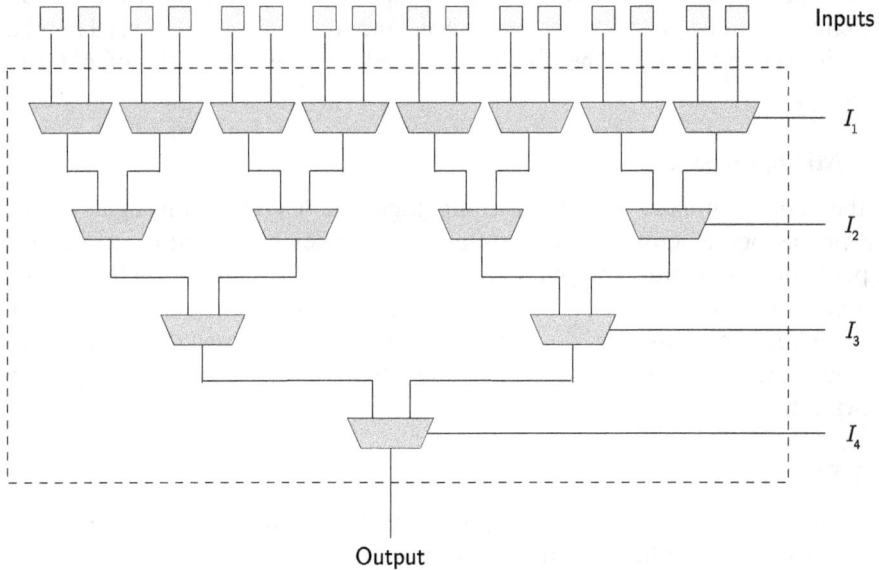

Figure A.2. A schematic for a 4-LUT made using 16 MUXs.

The multiplexer outputs one of the two remaining inputs based on the selection input's logic. A 4-LUT comprises four layers of 2-to-1 multiplexer (MUX) circuits. A 4-LUT has four selection inputs, shown as I_1, I_2, I_3, and I_4 in figure A.2, and 16 standard inputs connected to registers storing values of either 0 or 1.

Each logic block in the FPGA typically consists of a four-input[1] lookup table (4-LUT) and an optional D Flip-Flop. A 4-LUT can implement any logic gate if we alter the values stored in the input registers, and the D flip-flop halts the

---

[1] Some advanced FPGAs use LUTs with six control inputs as well.

output of the logic block until it registers a rising clock edge (we will get into the details of clock cycles further down the road).

A.3.2 Routing channels

A modern FPGA may contain tens of thousands of logic blocks. A web of routing wires and programmable interconnection switches interconnects all of these logic blocks. This combination allows us to interconnect all the logic blocks, if we desire so, to make the circuit we need.

A.3.3 I/O pads

Any circuit we make would be rendered useless if we could not communicate with it. For this purpose, FPGAs have I/O pads that enable them to establish input and output connections with various instruments and peripherals.

A microprocessor and an FPGA differ in how they are programmed. While a microprocessor has fixed hardware that needs a continuous input of commands it can process, an FPGA needs to be *configured* only once after power-up, the circuit gets implemented, and the required task can be performed as many times as needed. For this reason, programming an FPGA is often called *configuration*.

To configure an FPGA, we download a bit-stream file into it. The bit-stream file is a collection of 0s and 1s. It contains instructions from a hardware configuration code written in a HDL. The bit-stream file implements the required circuitry on the FPGA chip and stays implemented on the FPGA as long as it is powered on. There is absolutely no need to send further commands to the FPGA.

In research laboratories, scientists frequently develop new ideas and propose changes to the circuitry, or even the whole logical implementation. Hence, using FPGAs is ideal for research laboratories, including implementing quantum information processing and quantum communication tasks that employ photonic implementations. These are the precise applications that we focus on in this appendix.

A.4 Selecting a suitable FPGA

FPGAs are often embedded on development boards that complement the FPGA with some common and essential components. While FPGAs may seem to be the solution to all digital logic needs, there are some guidelines that one must consider before buying an FPGA development board. One FPGA cannot suit everyone's needs, and in the worst case one may buy something expensive yet unwieldy for the desired application. If you are building your quantum laboratory for the first time, you may get perplexed while choosing the right FPGA for your experimental setup. This section may help you navigate through this shopping journey.

A.4.1 Options for input and output

FPGA development boards are designed specifically for the industry they serve, especially in the input/output (I/O) department. An FPGA designed for use with multimedia devices or broadcasting purposes may have many multimedia connectors, such as video interface ports and audio jacks. On the other hand, the FPGA designed for use with some outer-space equipment may not contain any multimedia interface because no one needs audio jacks in outer space. Furthermore, I/O connectors will occupy some area on the FPGA board and will also add to the cost of the device.

A.4.2 Frequency

FPGA development boards are equipped with crystals that produce a digital signal oscillating between high and low voltages ('0' and '1') at a specific frequency. We call this pulsating voltage the clock signal. A rule of thumb to assess the speed of an FPGA development board is to check its frequency—the higher the frequency, the faster the device.

A.4.3 Cost

Depending on the budget, one may want to stay within the experimental needs and requirements. However, with a reasonable budget, a device with specifications to do a bit more than the currently required task might be a better choice, as it can be used if the experimental setup is expanded or improved. Factors that primarily impact the cost of a device include its clock speed, available logic units, number and variety of I/Os, and the size of RAM.

A.4.4 Manufacturer

Xilinx and Altera are currently the two major manufacturers of FPGAs. It is better to choose a reliable, trustworthy, and well-known producer because the community support is far wider and more accessible.

A.4.5 Our experimental needs and choice of FPGA

In our case, we use FPGAs to count pulses coming from photon detections. These pulses are registered every time the photon hits the detector. Additionally, in our quantum information experiments, we are often required to compare multiple pulses arising out of multiple events and count coincidences between two, three, or perhaps four channels. Everything is happening at the time scale of nanoseconds. So, the FPGA must be super fast.

The output of our optical setup after the photons hit the detector are electrical, square pulses with width 15 ns and peak-to-peak voltage of 2.2 V. For a typical experiment, we are being hit by hundreds of thousands of such pulses every second from up to four channels simultaneously. We humans are not fast nor sensitive enough to count all these pulses on our own and note them down on a piece of paper. Ordinary electronics may also fall short of keeping up with these requirements. Therefore, to meet our speed requirements and due to the dynamic nature of our research needs, we opt for an FPGA development board (figure A.3).

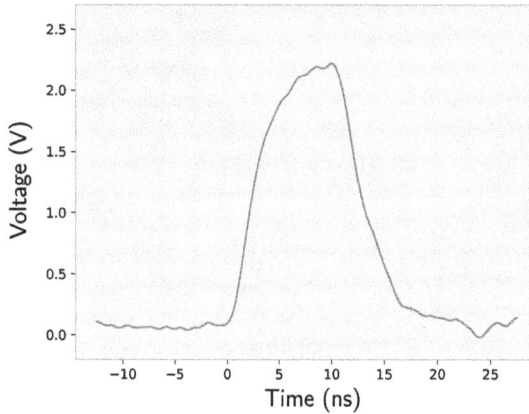

Figure A.3. Pulse from a detector.

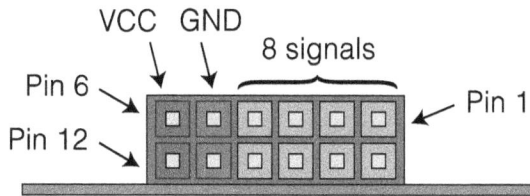

Figure A.4. Front view of a Pmod connector. Other than the ground pins, any pin can be used to collect input pulses.

Our CCU needs to detect and count the signal produced by our single-photon detectors, i.e., square pulses of about 15 ns. Our FPGA's clock speed should be such that it can register every count precisely without a miss. Theoretically, to detect a 20 ns pulse, we require the clock frequency of at least $1/(15 \times 10^{-9}) = 66.7$ MHz, while in practice it should be even higher. Furthermore, for experiments in the single-photon quantum laboratory, we will only require Digilent Pmod™ connectors (shown in figure A.4). Pins in Pmod connectors can be used as digital I/O communication pins to input the detector pulses.

The device we used for experiments was a Digilent™ Nexys A7 development board with a Xilinx Artix 7 FPGA embedded on it. It has four digital Pmod connectors and a crystal oscillator that produces a clock of 100 MHz. The I/O connections are more than sufficient and with our clock requirement of 66.7 MHz only, we would not be missing a beat with a 100 MHz clock at our disposal.

A.5 Overview of the circuitry

Before we jump into the coding phase, we should have a good idea of the functionality we need to implement. We need to develop a CCU that counts the incoming pulses from four channels (A, B, B', and A'), their coincidences, and sends the data over to the PC. The functionality of the CCU can be broken down into

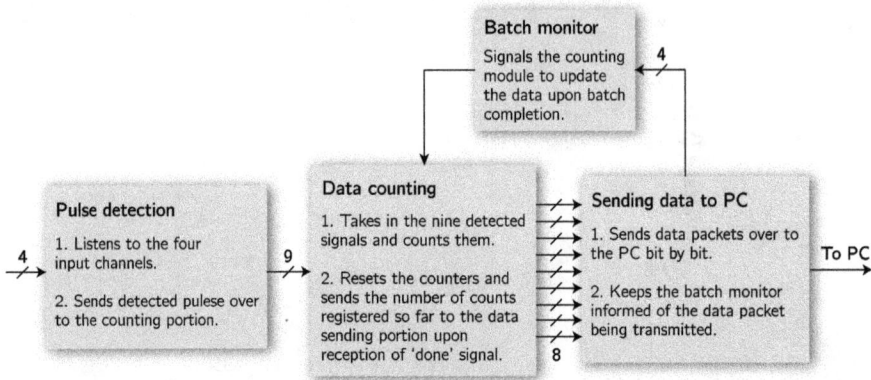

Figure A.5. A workflow chart summarizing the whole functionality of the CCU. Our implementation comprises a total of four major modules. These modules have connecting wires between them. The arrows in the figure show the direction of data flow, while the numbers over arrows with crosses represent the number of bits that each bus (a channel comprising more than 1 bit) carries. An arrow without a cross represents a single bit.

three major portions. The schematic workflow in figure A.5 shows how these portions are connected.

A.5.1 Pulse detection

In this portion, the circuit needs to generate the required single-pulse and coincidence pulse channels, detect the incoming pulses on these channels, and communicate each detection to the counting portion.

A.5.2 Data counting

This portion of our circuitry receives detection signals from the pulse detection and counts them until it gets the 'reset' signal. Once it does get the 'reset' signal, it stores the count value in the data-storage circuit elements called *registers* and starts counting the new pulses again from zero.

A.5.3 Send data to PC

This final portion sends the information about the pulses counted to the PC using a communication protocol known as UART, which stands for universal asynchronous receiver and transmitter.

Notice that the data counting and sending portions should be synchronized. The values available to the sending portion should change only when it has successfully transmitted one batch of data. The small portion of our circuitry, named the batch_monitor, ensures this synchronization by sending pulses that trigger the data update.

A.6 Writing code for FPGAs

FPGAs are built differently from microprocessors, and common programming languages such as C, Java, or Python that we use to program microprocessors will be of no use here. We have to implement circuit designs using logic gates and wires on FPGA. Therefore, in order to program an FPGA, we need a HDL to define physical connections between hardware resources available on the development board. We have two options to choose from: Verilog and VHDL. Our choice is Verilog, whose syntax resembles C, a famous programming language used with microprocessors. Therefore, using Verilog provides a heads-up for individuals with experience in C.

In Verilog, we divide the tasks we need to perform into portions of code called *modules*, and each module may be split further into hierarchically lower-level modules. The four portions of the flowchart in figure A.5 will depict the four core modules of our Verilog code, and they comprise several hierarchically lower modules. All the modules run simultaneously once the FPGA is configured. Unlike C or other programming languages, when we configure the FPGA there will be no program flow to dictate the sequence of execution of the code! However, we will need to define the interconnects between the modules to make sure the signals pass through the modules in the correct order to achieve the required functionality.

A.7 Programming the FPGAs

Once we have written HDL for our required functionality, we need to transfer this circuitry onto the FPGA in portions of codes we called modules. Before we plug in the wire and try to transfer the circuitry to the FPGA, there are some implications that we need to take care of first. How will our HDL implementation translate to a transistor-based circuit? And, how are we going to feed the single-photon pulses to a chip? In this section, we will address these concerns.

A.7.1 Defining physical connections

With the circuitry ready, we need to specify the pins or connectors on the development board to communicate to and from the FPGA. The assignment of a pin on the development board is known as a *constraint*. We use a single constraint file in the .xdc format to specify all the constraints for a project. Each pin in the FPGA board has a specific code as its name. The easiest way to specify the constraints is by using the master .xdc file for our development board[2]. All the lines in this file will be commented out by default. To set a constraint, we uncomment the line for the pin we intend to use and modify it a bit to connect it to our circuit's input/output wire.

Our circuitry requires four input pulses, one output pulse for data transfer to the PC, and a clock signal. For inputs, we use one of the Pmod connectors. For output, we use Nexys A7's onboard serial communication device, and the onboard oscillator

---

[2] GitHub repository for Digilent boards' master .xdc files: https://github.com/Digilent/digilent-xdc.

for the clock signal. The constraint file available in our GitHub repository contains information for the constraints we defined while using a Nexys A7.

A.7.2 Synthesizing and analyzing HDL

Once we have defined the constraints, we need an HDL synthesizing and analyzing software to convert our HDL implementation into gate-level circuits. While there are several options for such software available, we chose what our FPGA manufacturer provided, i.e., Xilinx Vivado. Vivado itself processes all the required steps, and we just need to give the commands.

In the *synthesis* step, Vivado determines the components needed to implement our circuit and how they should be routed, while in the *implementation* step it determines how it will physically connect the components in our specific FPGA development board. The output of the implementation step is a file known as the *netlist* saved in the .net format.

A.7.3 Generating the bit stream

As the last step before we can configure our circuitry onto the FPGA chip, we need to generate a bit-stream (.bit) file which, as the name suggests, is a list of 0s and 1s, encoding the information that the FPGA can understand. The FPGA will decipher this .bit file and make the commanded circuitry by connecting the available logic blocks and I/Os using the routing channels. To generate the bit stream, Vivado encodes the implemented circuitry from the netlist into bits that the FPGA can read and implement.

The steps can be executed one after another by simply clicking on *Run Synthesis*, *Run Implementation*, and *Generate Bit-stream* buttons in Vivado's flow navigator. Alternatively, we only click on the *Generate Bit-stream* button and it will go through synthesis and implementation steps on its own if required. Depending on the computer, the whole process may take 3–4 min.

Writing the code for the FPGA requires a lot of iterations and bench tests. We have already run all the required tests for the code available in our GitHub repository. Users who require the same functionality as we did can freely utilize this code alone without any need for alteration. The rest can refer to [5] and [6] to learn Verilog in depth and modify the code as they wish.

A.7.4 Configuring the FPGA

Finally, it is time to configure our circuit on the FPGA using the .bit file. Once the dialog for bit-stream generation pops up, we can open Vivado's built-in hardware manager. We attach the FPGA to the PC using a micro-USB to USB type-A connecting cable, connect to the FPGA from the hardware manager, and program the device using the .bit file Vivado just created. Once the code has finished configuring the FPGA, the *done* LED on the development board will light up, and we are all set to use the device in our experiments (figure A.6).

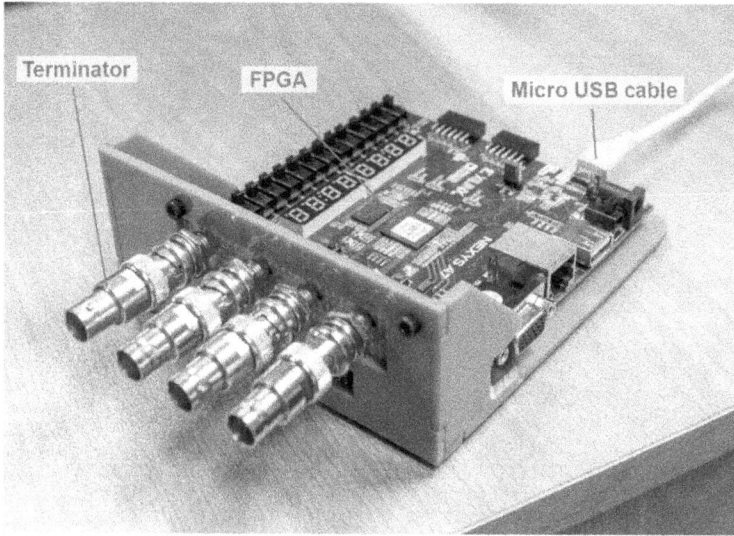

Figure A.6. Nexys A7 FPGA development board fitted on a custom housing, with a custom BNC connector bus and a micro-USB cable attached to it.

Figure A.7.

A.8 Reading data on a computer

For data logging, we have created a simple software called photo-detection counter (PDC) with an interactive Python GUI. Using this open-source programming language, we can communicate with the serial port to obtain data from the CCU. We have used Python's *pySerial* library to establish a serial communication port at the PC's end and *matplotlib* [4] library to create a graphical user interface (GUI) with real-time graphs (A.7).

The code for PDC is free to download from our website[3], and anyone can replicate or improve the code for similar experiments in single-photon laboratories. Figure A.8 shows a screenshot of the GUI.

The graphs on PDC show data for only a specified time window, and older counts are discarded to avoid congestion on the screen. The CAPTURE SCREEN button will take a screenshot of the window while the ACQUIRE DATA button will collect data for

---

[3] http://www.physlab.org/qmlab

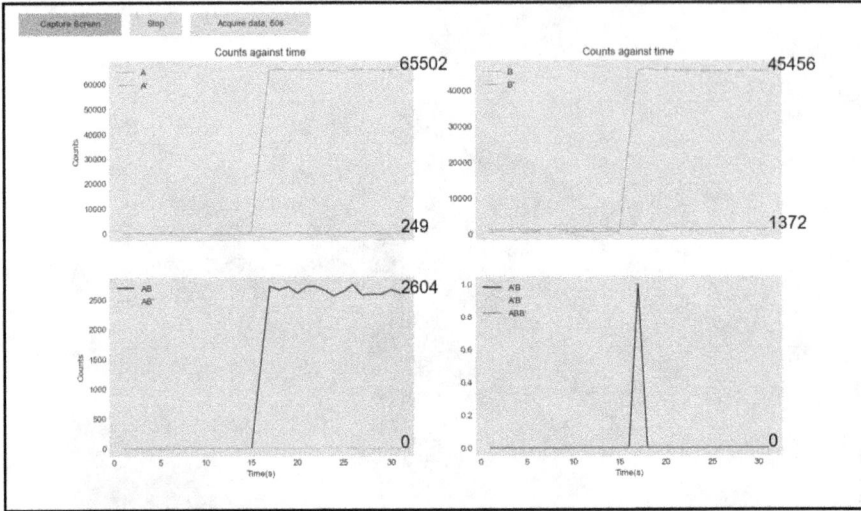

Figure A.8. GUI of the counting program written in Python.

the specified time and save it in `.txt` format. The data saved using ACQUIRE DATA will be saved with the following naming convention: 'data+*acquisition number*'. For example, if we keep collecting data in a single session, it will be saved as 'data0', 'data1', 'data2', and so on. All the data and screenshots will be saved in the same folder as the parent `.py` file. Finally, the STOP button securely closes the serial connection to the FPGA and terminates PDC.

When this Python script is run, it asks for the initial acquisition number, time period for each acquisition, display time window, baud rate, and communication port number for UART, and then launches the GUI.

A.9 Evaluating the system

Now that we have the CCU up and running, we need to evaluate it with some known input electronic pulses. This will help characterize and validate the system, and determine evaluation against benchmarks. We use Stanford Research system's delay generator, DG645, as our testing device[4]. Since we are designing a CCU specifically for single-photon laboratories, we will pay special attention to test cases where the simulated pulses from the DG645 resemble the pulses we are getting from our SPCMs, i.e., 2.2 V pulses of around 10 ns. For evaluation metrics, we specifically focus on communication rate, voltage input range, resolution, minimum and maximum counts, fidelity of coincidence, and integral nonlinearity (INL).

---

[4] DG645's specifications can be obtained from the following: link http://www.thinksrs.com/products/dg645.html.

A.9.1 Communication rate

As we saw in the previous sections, we need to transmit 10 packets of data in a batch where each packet is 10-bit long. Our baud rate for the serial connection with the PC is 4 MHz, so the data communication rate we will achieve is $(4 \times 10^6)/(10 \times 10) = 40\,000$ batches of counts every second. This means, theoretically, that our communication channel is able to inform us about single-photon detections individually for up to 40 000 photons per second. The maximum photon detection rate that we have achieved in our single-photon laboratory so far is about 20 000 single photons in a second, so this circuitry should be able to communicate each detection with high fidelity to the PC without taking a break.

A.9.2 Voltage input range

Digital signals work on logical highs and lows, which are voltage thresholds that digital devices see as 1s (highs) and 0s (lows). In this test, we determine the minimum voltage we need to provide to the FPGA input for it to register most of the pulses as a logical high. We send pulses of 20 ns to the FPGA at a known rate and vary the input voltage all the way from 0 V to 3.3 V (the voltage rating of FPGA input pins). Between the logical high and the logical low, we have the voltage region that is known as the forbidden region where the digital device may or may not register the pulse. The maximum voltage that can be supplied to the FPGA can be found in the manufacturer's specification sheet. For Nexys A7, a voltage greater than 5 V can permanently damage the device and should never be fed through the input pins. While testing with 10 ns wide pulses being sent at a rate of 10 000 pulses/s and voltage increasing in steps of 0.01 V, the counts started registering at 1.49 V, but this was the forbidden region. We increased the voltage until the change in counts per step was nearly zero. This is the saturation region, and while testing it with our circuitry it came out to be 1.6 V.

A.9.3 Resolution

The resolution of our counter refers to the minimum width of the pulse—above the detection threshold voltage—that it can detect. We use 2.2 V pulses at a rate of 10 000 pulses/s. Starting with an initial pulse width of 1 ns, we gradually decrease the width in steps of 1 ns. There are two metrics that we use to evaluate the resolution. The first is the minimum pulse width when the FPGA starts registering counts and the second is the minimum width required to reach the saturation point, just like we did to determine the voltage range. While performing the test, the FPGA started registering the pulses at 6 ns, but the saturation point was reached at the pulse width of 10 ns.

A.9.4 Coincidence window

The coincidence window is the main concern for the experiments performed in this book. It is used frequently to determine the accidental counts in single-photon experiments. The smaller the coincidence window, the lower the accidental counts

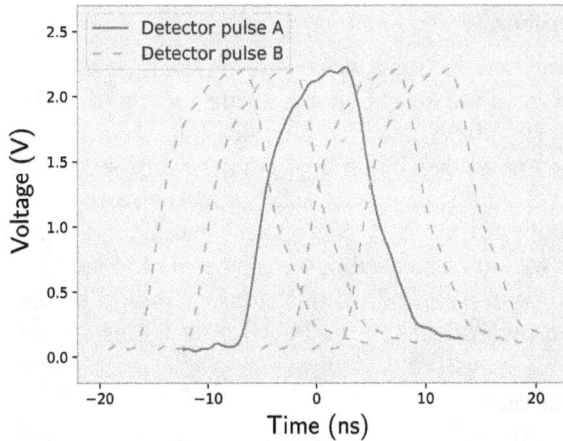

Figure A.9. This figure shows one of the pulses (dashed orange) being delayed with respect to the other. When the orange pulse is toward the left, the delay is negative. When on the right, the delay is positive.

will be, and the better will be our chance of detecting only the simultaneously released downconverted photons. Our detector's specifications sheet shows that it produces 2.2 V pulses of 10 ns pulse width. So, to determine the effective time window required to register a coincidence, we generate square pulses of 2.2 V and 10 ns pulse width at a rate of 10 000 pulses/s. Initially, we lag one pulse behind the other by 15 ns. We will denote this lag with a negative sign, so our initial delay is −15 ns. After this, we add a delay between the two channels in steps of 1 ns until the delay between the pulses is +15 ns. We will pay special attention to the saturation region in this case. We will refer to the time window for which the counts stay in the saturation region, despite there being a delay in the two pulses, as the coincidence time window. Figure A.9 gives us a visual representation of the delay between the two channels.

A.9.5 Minimum and maximum counts

We supplied the FPGA with square pulses of width 10 ns and 2.2 V peak voltage—both of these parameters being well within the saturation region, as we have already seen in the previous sections—to determine the count range. We started off with a minimum count rate of 1 count/s and increased it until the FPGA's output count rate saturated. Since we use 8-bit registers to store the values, the maximum number we can store in a register is $2^8 - 1 = 255$. Each registered value is sent to the PC 40 000 times every second, so the maximum count rate we can expect is,

$$\text{counts}_{\text{max}} = 255 \times 40\,000$$
$$= 1.02 \times 10^6 \text{ counts/s}$$

This maximum value may differ in the actual circuit implementation (accounting for propagation delays, clock skews, etc). Our main objective for this test is to ensure that the FPGA can handle the count rate we are sending to it.

A.9.6 Fidelity of coincidences

Coincidence counting is the most important attribute for single-photon experiments. The CCU should register the required coincidences for all simultaneous pulses arriving at the input channels. Fidelity is a numerical measure of this coincidence-detection ability and can be calculated using the following formula,

$$\mathcal{F} = \frac{\text{Detected coincidences}}{\text{Total simultaneous pulses}}.$$

We have implemented nine counters in our current FPGA circuitry, so we need to evaluate $\mathcal{F}$ for each of these individual counters.

A.9.7 Integral nonlinearity

An ideal counter would count each pulse perfectly without any missed or extra counts. In this case, the graph of counted pulses against pulses sent would give us a straight line with a gradient one. However, these counters may under-count or over-count these pulses in practical implementations. In many cases, this error gets *integrated*, i.e., the error keeps on increasing for increasing counts. The effect of this error can be visualized as a shift in the ideal straight line with gradient one, hence the name *integral nonlinearity* (figure A.10).

   We will evaluate these erroneous counts as a fraction of the expected counts using the following formula,

$$\mathcal{I} = \frac{\text{Measured counts} - \text{Actual counts}}{\text{Actual counts}}$$

Just like fidelity, we will have to evaluate the INL for each of the nine counters individually.

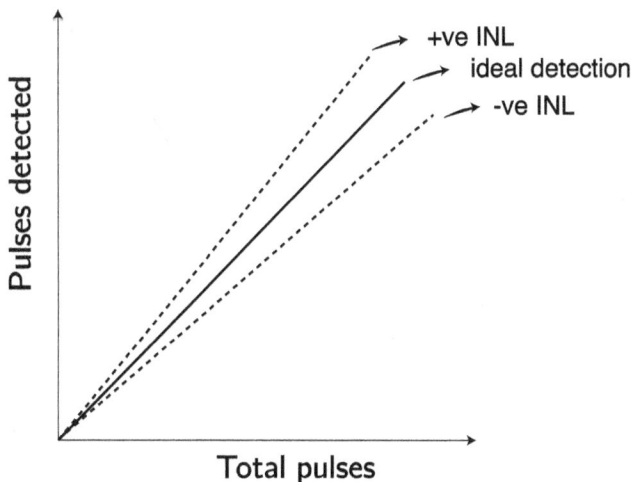

Figure A.10. Comparison of an ideal detection curve (solid line), with those containing an INL (dashed lines).

A.9.8 Evaluation summary

The table below summarizes the data for the CCU circuit programmed on a Nexys A7 FPGA development board.

Table A.1. This table summarizes the evaluation of CCU discussed in this section.

Maximum counts	$\approx 8 \times 10^6$
Resolution	6 ns
Voltage detection range	1.6–4.9 V
Coincidence window	3 ns

This CCU offers counting for up to 8 million photons per second, which is more than enough for our experiments in which our source outputs a maximum of 20 000 photon counts per second. Furthermore, our detectors output 2.2 V pulses that are 10 ns wide. These values are above our minimum ranges of both voltage and pulse width, so this CCU can detect the detector pulses without significant losses. Finally, the coincidence window of 3 ns is a significant improvement over our previous 40 ns coincidence window. This will make experiments more robust and less prone to accidental photodetections.

We are dealing with a total of nine counters in our circuit implementation, and we associate a distinct INL and fidelity with each of these counters. Table A.2 contains

Table A.2. This table lists down the INL and coincidence counting fidelity associated with each output of our CCU.

Output	A	B	B'	A'	AB	AB'	$A'B$	$A'B'$	ABB'
$\mathcal{I}(10^{-3})$	25.6	9.6	13.6	29.7	30.2	11.5	16.6	−5.0	5.1
$\mathcal{F}$	N/A	N/A	N/A	N/A	0.93	0.95	0.94	0.95	0.94

the data for INL and fidelity of each counter.

These values may differ for every new circuitry, even when designed on the same board.

In this appendix, we have discussed the development of a CCU on the Nexus A7 development board. FPGA based CCUs are inexpensive alternatives for expensive units available in the market. The code for this CCU is open source and can be downloaded from our website[5] or GitHub[6] repository.

---

[5] https://physlab.org/qmlab
[6] https://github.com/bilalshah07/coincidence_counting_fpga

References

[1] Branning D and Beck M 2012 An FPGA-based module for multiphoton coincidence counting *Advanced Photon Counting Techniques VI* **vol 8375** (Bellingham, WA: International Society for Optics and Photonics) p 83750F

[2] Li W, Hu Y, Zhong H, Wang Y, Wang X, Peng C and Jiang X 2018 Time-tagged coincidence counting unit for large-scale photonic quantum computing *Rev. Sci. Ins.* **89** 103113

[3] Hloušek J, Grygar J, Dudka M and Ježek M 2024 High-resolution coincidence counting system for large-scale photonics applications *Phys. Rev. Appl.* **21** 024023

[4] Hunter J D 2007 Matplotlib: A 2D graphics environment *CiSE* **9** 90–5

[5] Ciletti M D 2010 *Advanced Digital Design with the Verilog HDL* 2nd edn (Beijing: Publishing House of Electronics Industry)

[6] Roth B K L C and John L K 2015 *Digital Systems Design Using Verilog* 1st edn (Boston, MA: Cengage)

IOP Publishing

Quantum Mechanics in the Single-Photon Laboratory
(Second Edition)

Muhammad Sabieh Anwar, Faizan-e-Ilahi, Syed Bilal Hyder and Muhammad Hamza Waseem

Appendix B

Inventory for single-photon experiments

Optical elements

Item	Quantity	Company	Part number
Laser-405 nm	1	CNI Laser	MDL-III-405-50 mW
Mirror	2	Thorlabs	PF05-03-P01
HWP-405 nm	1	Thorlabs	WPH05M-405
BBO	1	Newlight Photonics	PABBO5050-405(1)
HWP-808 nm	2	Thorlabs	WPH05M-808
QWP-808 nm	2	Thorlabs	WPQ05M-808
PBS	2	Thorlabs	PBS252
BDP	2	Thorlabs	BD40
Polarizer	1	Thorlabs	GTH5M
Quartz crystal	1	MTI Corporation	SOX101005S2
Beam block	1	Thorlabs	LB1
Alignment Laser	1	Thorlabs	CPS180
Filters	4	Thorlabs	FGL780
Lens/collimators	5	Thorlabs	F220FC-780
Silica gel			

Mechanical elements

Item	Quantity	Company	Part number
Optical table	1	Thorlabs	T1220CK
Kinematic mounts for mirrors	2	Thorlabs	KM100
Kinematic mounts for BBO	1	Thorlabs	KM100
Motorized rotation mounts for wave plates	6	Thorlabs	PRM1Z8
Rotation mounts for polarizer	1	Thorlabs	RSP1(/M)
Kinematic prism mounts for PBS	2	Thorlabs	KM100PM/M
Kinematic mounts for BDP	2	Thorlabs	KM100
Lens/collimator adapter	5	Thorlabs	AD15F
Kinematic mounts for lens/collimator	5	Thorlabs	KC1/M
Fiber-fiber couplers	2	Thorlabs	FCB1
Post holder base		Thorlabs	BA1/M and BA2(/M)
Screws			M6
Posts		Homemade	
Post holders		Homemade	

Actuators and controllers

Item	Quantity	Company	Part number
DC servo motor controller for motorized rotation stage	4	Thorlabs	KDC101
Piezo inertia actuator for BDP mount	1	Thorlabs	PIAK10
Piezo actuator controller	1	Thorlabs	KIM101

Detection and coincidence counting unit

Item	Quantity	Company	Part number
Detectors	4	Excelitas Technologies	SPCM-AQRH
FPGA development board	1	Digilent	Nexys A7
Power supplies	4	Thorlabs	LDS5
Optical fibers	10	Thorlabs	M64L01
FPGA-APD connector	1	Homemade	
BNC cables			
50 Ohm BNC terminators	4		
Computer with Python	1		

Testing of FPGA

Item	Quantity	Company	Part number
FPGA development board	1	Digilent	Nexys A7
Delay generator	1	Stanford Research Systems	DB64
Pulse generator	1	Stanford Research Systems	DG645
High speed oscilloscope	1	Agilent	DSO6104L
BNC cables			

These tables incorporate the complete inventory for the laboratory. Below, we tabulate individual lists for each quantum experiment discussed in this book. In these lists, for the sake of brevity, we do not include obvious components such as an optical table and computer or mounts and posts.

Q1: Spontaneous parametric downconversion

Item	Quantity	Company	Part number
Laser-405 nm	1	CNI Laser	MDL-III-405-50 mW
Mirror	2	Thorlabs	PF05-03-P01
HWP-405 nm	1	Thorlabs	WPH05M-405
BBO	1	Newlight Photonics	PABBO5050-405(1)
Beam block	1	Thorlabs	LB1
Detector	2	Excelitas Technologies	SPCM-AQRH

Q2: Proof of existence of photons

Item	Quantity	Company	Part number
Laser-405 nm	1	CNI Laser	MDL-III-405-50 mW
Mirror	2	Thorlabs	PF05-03-P01
HWP-405 nm	1	Thorlabs	WPH05M-405
BBO	1	Newlight Photonics	PABBO5050-405(1)
Beam block	1	Thorlabs	LB1
HWP-808 nm	1	Thorlabs	WPH05M-808
PBS	1	Thorlabs	PBS252
Delay generator	1	Stanford Research Systems	DB64
Detector	3	Excelitas Technologies	SPCM-AQRH

Q3: Estimating the polarization state of single photons

Item	Quantity	Company	Part number
Laser-405 nm	1	CNI Laser	MDL-III-405-50 mW
Mirror	2	Thorlabs	PF05-03-P01
HWP-405 nm	1	Thorlabs	WPH05M-405
BBO	1	Newlight Photonics	PABBO5050-405(1)
Beam block	1	Thorlabs	LB1
HWP-808 nm	2	Thorlabs	WPH05M-808
QWP-808 nm	2	Thorlabs	WPQ05M-808
PBS	1	Thorlabs	PBS252
Detector	3	Excelitas Technologies	SPCM-AQRH

Q4: Visualizing the polarization state of single photons

Item	Quantity	Company	Part number
Laser-405 nm	1	CNI Laser	MDL-III-405-50 mW
Mirror	2	Thorlabs	PF05-03-P01
HWP-405 nm	1	Thorlabs	WPH05M-405
BBO	1	Newlight Photonics	PABBO5050-405(1)
Beam block	1	Thorlabs	LB1
HWP-808 nm	1	Thorlabs	WPH05M-808
QWP-808 nm	2	Thorlabs	WPQ05M-808
Polarizer	1	Thorlabs	GTH5M
Detector	2	Excelitas Technologies	SPCM-AQRH

NL1: Freedman's test of local realism

Item	Quantity	Company	Part number
Laser-405 nm	1	CNI Laser	MDL-III-405-50 mW
Mirror	2	Thorlabs	PF05-03-P01
HWP-405 nm	1	Thorlabs	WPH05M-405
Quartz crystal	1	MTI Corporation	SOX101005S2
BBO	1	Newlight Photonics	PABBO5050-405(1)
Beam block	1	Thorlabs	LB1
Polarizer	2	Thorlabs	LPNIRE050-B
Detector	2	Excelitas Technologies	SPCM-AQRH

NL2: Hardy's test of local realism

Item	Quantity	Company	Part number
Laser-405 nm	1	CNI Laser	MDL-III-405-50 mW
Mirror	2	Thorlabs	PF05-03-P01
HWP-405 nm	1	Thorlabs	WPH05M-405
Quartz crystal	1	MTI Corporation	SOX101005S2
BBO	1	Newlight Photonics	PABBO5050-405(1)
Beam block	1	Thorlabs	LB1
HWP-808 nm	2	Thorlabs	WPH05M-808
PBS	2	Thorlabs	PBS252
Detector	4	Excelitas Technologies	SPCM-AQRH

NL3: CHSH test of local realism

Item	Quantity	Company	Part number
Laser-405 nm	1	CNI Laser	MDL-III-405-50 mW
Mirror	2	Thorlabs	PF05-03-P01
HWP-405 nm	1	Thorlabs	WPH05M-405
Quartz crystal	1	MTI Corporation	SOX101005S2
BBO	1	Newlight Photonics	PABBO5050-405(1)
Beam block	1	Thorlabs	LB1
HWP-808 nm	2	Thorlabs	WPH05M-808
PBS	2	Thorlabs	PBS252
Detector	4	Excelitas Technologies	SPCM-AQRH

Q5: Single-photon interference and quantum erasure

Item	Quantity	Company	Part number
Laser-405 nm	1	CNI Laser	MDL-III-405-50 mW
Mirror	2	Thorlabs	PF05-03-P01
HWP-405 nm	1	Thorlabs	WPH05M-405
BBO	1	Newlight Photonics	PABBO5050-405(1)
Beam block	1	Thorlabs	LB1
HWP-808 nm	2	Thorlabs	WPH05M-808
BDP	2	Thorlabs	BD40
Polarizer	1	Thorlabs	LPNIRE050-B
Detector	2	Excelitas Technologies	SPCM-AQRH

Q+NL: Nonlocal quantum erasure

Item	Quantity	Company	Part number
Laser-405 nm	1	CNI Laser	MDL-III-405-50 mW
Mirror	2	Thorlabs	PF05-03-P01
HWP-405 nm	1	Thorlabs	WPH05M-405
BBO	1	Newlight Photonics	PABBO5050-405(1)
Beam block	1	Thorlabs	LB1
HWP-808 nm	3	Thorlabs	WPH05M-808
PBS	2	Thorlabs	PBS252
BDP	2	Thorlabs	BD40
Polarizer	1	Thorlabs	LPNIRE050-B
Detector	3	Excelitas Technologies	SPCM-AQRH

QST: Quantum state tomography

Item	Quantity	Company	Part number
Laser-405 nm	1	CNI Laser	MDL-III-405-50 mW
Mirror	2	Thorlabs	PF05-03-P01
HWP-405 nm	1	Thorlabs	WPH05M-405
Quartz crystal	1	MTI Corporation	SOX101005S2
BBO	1	Newlight Photonics	PABBO5050-405(1)
Beam block	1	Thorlabs	LB1
QWP-808 nm	2	Thorlabs	WPQ05M-808
HWP-808 nm	2	Thorlabs	WPH05M-808
PBS	2	Thorlabs	PBS252
Detector	4	Excelitas Technologies	SPCM-AQRH

www.ingramcontent.com/pod-product-compliance
Lightning Source LLC
Chambersburg PA
CBHW080542220326
41599CB00032B/6335